PASCAL PROGRAMS
for Scientists
and
Engineers

PASCAL PROGRAMS
for Scientists
and
Engineers

Alan R. Miller
Professor of Metallurgy
New Mexico Institute of Mining and Technology

CREDITS
Cover Design by Daniel Le Noury
Technical illustrations by J. Trujillo Smith, Jeanne E. Tennant

Library of Congress Card Number: 81-51128
ISBN 089588-058-X
First Edition 1981
Printed in the United States of America
10 9 8 7 6 5 4

Contents

4   *Simultaneous Solution of Linear Equations*   **59**

5   *Development of a Curve-Fitting Program*   **137**

Appendix A: Reserved Words and Functions   353

Appendix B: Summary of Pascal   355

Bibliography   371

Index   373

Listings and Illustrations

Chapter 8

Chapter 9

Preface

The ideas and material for this book have been developed during my experience teaching sophomore, junior, and senior engineering students. Over the past 14 years the computer language I have used for these classes has changed from FORTRAN to BASIC to what is perhaps the ideal language for the classroom—Pascal.

All of the Pascal programs in this book were developed on a Z-80 microcomputer. The operating system was the Lifeboat 2.2 version of CP/M. The source programs were written with MicroPro's Word-Master text editor and compiled with Pascal/M. Most of the programs were also run on other Pascal compilers. These included the CP/M versions of Pascal/MT+, Pascal/Z, and JRT Pascal, and a Pascal compiler running on a Dec-20.

The manuscript was created and edited with MicroPro's WordStar running on the same Z-80 computer. The Pascal source programs have been incorporated directly into the manuscript from the original source files. Computer printouts shown in the figures were also incorporated magnetically into the manuscript. This was accomplished by altering the CP/M operating system so that printer output was written into a block of memory. The final manuscript was submitted to SYBEX in a magnetic form compatible with the photocomposer. Consequently, the manuscript and the Pascal source programs have not been retyped.

I am sincerely grateful for the helpful guidance and suggestions of Rudolph Langer and Douglas Hergert during the development of the manuscript. I would also like to thank Ashok Singh for checking the manuscript, especially the mathematical expressions.

Alan R. Miller
Socorro, New Mexico
May 1981

Introduction

The purpose of this book is twofold: to help the reader develop a proficiency in the use of Pascal, and to build a library of programs that can be used to solve problems frequently encountered in science and engineering.

The programs in this book will prove valuable to the practicing scientist or engineer. The material is also suitable for a junior- or senior-level engineering course in numerical methods. The reader should have a working knowledge of an applications language such as Pascal, FORTRAN, or BASIC. In addition, experience with vector operations and with differential and integral calculus will be helpful.

The many distinctive features of Pascal lead to clear and elegant programming practices. Two of these features that are particularly valuable in scientific programs are *long variable names* and *block structure*. When identifiers such as SUM_X_SQUARED and NO_CHANGE can be utilized in preference to shorter symbols, the resulting source code is easier to understand. Pascal also offers a greater variety of iterative statements than those supplied in the most common BASIC and FORTRAN compilers. The statement

FOR I := 1 **TO** N **DO**

corresponds to the **FOR ... NEXT** loop in BASIC and **DO ... CONTINUE** in FORTRAN. In Pascal this statement is complemented by the other iterative constructions **WHILE ... DO** and **REPEAT ... UNTIL**.

The major features of Pascal are summarized in *Pascal User Manual and Report*, by Jensen and Wirth[1]; additional details can be found in *Introduction to Pascal*, by Zaks[2]. However, not all of the features included in the original specification of Pascal have been incorporated into actual Pascal compilers. For example, most Pascal compilers do not allow

---

1. Kathleen Jensen and Niklaus Wirth, *Pascal User Manual and Report*, Second Edition (New York, Heidelberg, and Berlin: Springer-Verlag, 1974).

2. Rodnay Zaks, *Introduction to Pascal, Including UCSD Pascal* (Berkeley: Sybex, 1980).

procedure and function names to be used as parameters to other procedures. (This problem is discussed in Chapter 8 under the heading "Generalizing Procedure Calls.")

On the other hand, several desirable features, omitted from the original definition of Pascal, have been added to commercial Pascal compilers. (Unfortunately, these additions have not been uniformly implemented.) One such feature is the incorporation of the dynamic string type. Another is the inclusion of a random number generator, a feature discussed in Chapter 2.

Throughout this work, features common to other high-level languages such as FORTRAN, BASIC, and ALGOL have frequently been utilized in preference to the more elegant techniques of Pascal. For example, arrays are used in preference to records. As a consequence, the algorithms presented here can be more easily converted into other languages.

The reader who is primarily interested in the Pascal programs developed in this book will have no trouble locating them; the sections that contain programs, procedures, or functions are clearly labeled. However, this book is designed to be read from beginning to end. Each chapter discusses and develops tools that will be used again in subsequent chapters. The mathematical algorithms of each program are methodically described before the program itself is implemented, and sample output is supplied for most of the programs. The following brief descriptions summarize the contents of each chapter:

Chapter 1, **Evaluation of a Pascal Compiler**, identifies weak points in several commercial compilers, and supplies programs for testing any Pascal compiler. The results of these tests will be used to select various constants and procedures in later chapters. Also included in Chapter 1 is a discussion of logarithmic, exponential, and trigonometric functions in Pascal.

Chapter 2, **Mean and Standard Deviation**, discusses some basic statistical algorithms and presents a program for implementing them. Procedures for generating—and testing—both uniform and Gaussian random numbers are also given.

Chapter 3, **Vector and Matrix Operations**, summarizes the operations of vector and matrix arithmetic, including dot product, cross product, matrix multiplication, and matrix inversion. Two important programs are developed—one for carrying out matrix multiplication, and another for calculating determinants.

Chapter 4, **Simultaneous Solution of Linear Equations**, presents programs to carry out the algorithms of Cramer's rule, the Gauss elimination method, the Gauss-Jordan elimination method, and the Gauss-Seidel method—all for solving simultaneous equations. In addition, *ill-conditioning* is studied through a program that generates Hilbert matrices, and a program is developed for solving equations with complex coefficients.

Chapter 5, **Development of a Curve-Fitting Program**, is the first of a series of chapters on curve fitting. In a good illustration of top-down program development, a linear least-squares curve-fitting program is written and discussed. The program includes procedures to simulate data, plot curves, compute the fitted curve, and supply the correlation coefficient.

Chapter 6, **Sorting**, describes and compares several Pascal sorting routines, including two bubble sorts, a Shell sort, and a recursive and nonrecursive quick sort. A sort routine is incorporated into the curve-fitting program of Chapter 5 to enable the program to handle real experimental data.

Chapter 7, **General Least-Squares Curve Fitting**, extends the curve-fitting program to general polynomial equations, and finds curve fits for three examples: heat capacity, vapor pressure, and superheated steam.

Chapter 8, **Solution of Equations by Newton's Method**, presents a series of programs that use Newton's algorithm for finding the roots of an equation. This tool will be used again in Chapter 10 for nonlinear curve fitting.

Chapter 9, **Numerical Integration**, develops programs for three different integration methods—trapezoidal rule, Simpson's rule, and the Romberg method. *End correction* is also discussed. Simpson's rule will be used in Chapter 11 for evaluating the Gaussian error function.

Chapter 10, **Nonlinear Curve-Fitting Equations**, discusses curve-fitting algorithms for the rational function and the exponential function. Two examples are given—the Clausing factor, and the diffusion equation.

Chapter 11, **Advanced Applications: The Normal Curve, the Gaussian Error Function, the Gamma Function, and the Bessel Function**, addresses several advanced topics in programming for mathematical applications. This last chapter summarizes and expands upon a number of the concepts presented earlier in the book.

For the reader who is approaching Pascal for the first time, a summary of the syntax, standard functions, and reserved words of Pascal is included in the appendices. It is hoped, however, that the real educational experience of this book will be gained through a careful study of the programs themselves.

A Note on Typography

The text of this book has been set in the typeface known as Oracle; the programs are in Futura, and the output has been photographed from the actual line-printer output supplied by the author. Pascal reserved words appear in **boldface**. Mathematical expressions (variables, letter constants) appear in *italics*, except for vectors, which appear in boldface roman. For example:

$$A + Bx + Cx^2 = 0$$

and

$$\mathbf{v} = \begin{bmatrix} 1 & 2 & 3 \end{bmatrix}$$

An effort has been made throughout the book to distinguish typographically between mathematical values and program structures. Thus, juxtaposed in a single paragraph the reader may see references to the variable x and the Pascal scalar X; the vector $\mathbf{v}$ and the Pascal array V; the matrix element v_{ij} and the Pascal array element V[I,J].

CHAPTER **1**

Evaluation of a Pascal Compiler

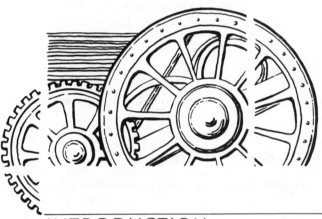

INTRODUCTION

To understand the results of a Pascal program we must be familiar with the limitations of the compiler we are using. This is particularly true with scientific applications programs such as the ones given in this book. In this first chapter, then, we will present some tools for evaluating the precision and range of any Pascal compiler. In fact, the examples given here were derived from several commercially available Pascal compilers.

In our evaluation we will consider a number of Pascal functions—logarithmic, exponential, and trigonometric. By this means we will see some of the inadequacies of actual compilers. We will explore ways of testing Pascal functions and ways of deriving functions that do not exist on typical Pascal compilers.

PRECISION AND RANGE OF FLOATING-POINT OPERATIONS

Several programs in this book are sensitive to the *precision* and *dynamic range* of the Pascal floating-point operations. For example, in one program an algorithm is terminated when a particular term is smaller than a relative tolerance. The formula in this case is:

$$\text{TERM} < \text{TOL} * \text{SUM}$$

where TERM is the value of the new term, SUM is the current total, and TOL is an arbitrarily small number known as the *tolerance*.

It is important that the value chosen for the tolerance not be outside the accuracy of the floating-point operations. Otherwise, the summation step will never terminate. Suppose, for example, that the floating-point operations are performed to a precision of six significant figures. Then, the tolerance must be set to a value that is larger than 10^{-6}.

The dynamic range of the exponent is a separate matter. Typical binary, floating-point operations are performed with 32 bits of precision. BCD floating-point packages, on the other hand, will usually have a greater dynamic range.

In the following section we will present a Pascal program that tests the precision and range of a compiler. We will investigate output from several compilers to illustrate both mantissa and exponent accuracy.

PASCAL PROGRAM:
A TEST OF THE FLOATING-POINT OPERATIONS

The program given in Figure 1.1 can be used to determine the precision and the dynamic range of a Pascal compiler. Type up the program and execute it. The initial value of X is obtained by dividing 10^{-4} by 3. Then, successively smaller and smaller values of X are calculated and displayed on the console. Each succeeding value is obtained from the product of 0.1 and the previous value. The process continues until 40 values have been printed or until a floating-point error terminates the program.

Let us now use this program to test the accuracy and range of three different compilers.

Three Runs of the Program: A Comparison

The initial mantissa is chosen to be ⅓, a repeating fraction that cannot be precisely represented by a floating-point number. Successive multiplications will show the extent of roundoff error. A 32-bit binary, floating-point number typically utilizes three bytes for the mantissa and

```
PROGRAM TEST(OUTPUT);
(* test range of floating point numbers *)

VAR
    I : INTEGER;
    X : REAL;

BEGIN
    WRITELN;
    X := 1.0E − 4/3.0;
    FOR I := 1 TO 40 DO
        BEGIN
            WRITE(' x =', X);
            X := 0.1 * X;
            WRITELN('   x =', X);
            X := 0.1 * X
        END
END.
```

Figure 1.1: A Test of the Floating-Point Operations

one byte for the exponent. This usually produces six or seven significant figures of precision and a dynamic range of 10^{+38} to 10^{-38}. The result might look like the output in Figure 1.2. In this first example, a floating-point error terminated the program at the limit of the dynamic range. Accuracy of the mantissa at this point is about five significant figures. The effect of roundoff errors is apparent in the least-significant digits.

As another example, consider the output (shown in Figure 1.3) from a different Pascal compiler that also uses a 32-bit binary floating-point number. In this case, the mantissa shows a little less precision. But more importantly, the dynamic range is only 10^{-18}. Furthermore, the floating-point underflow did not terminate the program and so the results became meaningless.

The output from a third Pascal compiler that uses 64-bit, BCD floating-point numbers is shown in Figure 1.4. The mantissa contains 14 digits of precision and the dynamic range is 10^{+63} to 10^{-63}. Notice that in this example, as in the previous one, the computation was not terminated by a floating-point error.

```
x = 3.33333E-5    x = 3.33333E-6
x = 3.33333E-7    x = 3.33333E-8
x = 3.33333E-9    x = 3.33333E-10
x = 3.33333E-11   x = 3.33333E-12
x = 3.33333E-13   x = 3.33333E-14
x = 3.33333E-15   x = 3.33333E-16
x = 3.33333E-17   x = 3.33333E-18
x = 3.33333E-19   x = 3.33333E-20
x = 3.33333E-21   x = 3.33333E-22
x = 3.33333E-23   x = 3.33333E-24
x = 3.33333E-25   x = 3.33333E-26
x = 3.33333E-27   x = 3.33334E-28
x = 3.33334E-29   x = 3.33334E-30
x = 3.33334E-31   x = 3.33334E-32
x = 3.33334E-33   x = 3.33334E-34
x = 3.33334E-35   x = 3.33334E-36
x = 3.33334E-37   x = 3.33334E-38
x =
Floating point error      (error message from compiler)
```

Figure 1.2: Precision Test: First Compiler

```
x = .3333334E-04   x = .3333333E-05
x = .3333332E-06   x = .3333330E-07
x = .3333330E-08   x = .3333329E-09
x = .3333328E-10   x = .3333327E-11
x = .3333327E-12   x = .3333325E-13
x = .3333325E-14   x = .3333325E-15
x = .3333323E-16   x = .3333323E-17
x = .3333323E-18   x = .0000000E-18
x = .1229780E+19   x = .1134271E+18
x = .1134271E+17   x = .1134271E+16
• • •
```

Figure 1.3: Precision Test: Second Compiler

A rather interesting bug contained in several compilers can limit the range of the Pascal SIN and COS functions. We will now investigate this phenomenon.

PASCAL SIN AND COS FUNCTIONS

A problem can occur with the SIN and COS functions as the argument approaches zero. When the magnitude of the argument is less than 10^{-8} or so, the SIN function should return the argument and the COS function should return unity. But several commercial Pascal compilers contain an error. The argument is squared before a range check is performed. This produces a floating-point underflow and meaningless results.

```
x  =  +0.33333333333333E-04    x  =  +0.33333333333333E-05
x  =  +0.33333333333333E-06    x  =  +0.33333333333333E-07
x  =  +0.33333333333333E-08    x  =  +0.33333333333333E-09
x  =  +0.33333333333333E-10    x  =  +0.33333333333333E-11
x  =  +0.33333333333333E-12    x  =  +0.33333333333333E-13
x  =  +0.33333333333333E-14    x  =  +0.33333333333333E-15
x  =  +0.33333333333333E-16    x  =  +0.33333333333333E-17
x  =  +0.33333333333333E-18    x  =  +0.33333333333333E-19
x  =  +0.33333333333333E-20    x  =  +0.33333333333333E-21
x  =  +0.33333333333333E-22    x  =  +0.33333333333333E-23
x  =  +0.33333333333333E-24    x  =  +0.33333333333333E-25
x  =  +0.33333333333333E-26    x  =  +0.33333333333333E-27
x  =  +0.33333333333333E-28    x  =  +0.33333333333333E-29
x  =  +0.33333333333333E-30    x  =  +0.33333333333333E-31
x  =  +0.33333333333333E-32    x  =  +0.33333333333333E-33
x  =  +0.33333333333333E-34    x  =  +0.33333333333333E-35
x  =  +0.33333333333333E-36    x  =  +0.33333333333333E-37
x  =  +0.33333333333333E-38    x  =  +0.33333333333333E-39
x  =  +0.33333333333333E-40    x  =  +0.33333333333333E-41
x  =  +0.33333333333333E-42    x  =  +0.33333333333333E-43
x  =  +0.33333333333333E-44    x  =  +0.33333333333333E-45
x  =  +0.33333333333333E-46    x  =  +0.33333333333333E-47
x  =  +0.33333333333333E-48    x  =  +0.33333333333333E-49
x  =  +0.33333333333333E-50    x  =  +0.33333333333333E-51
x  =  +0.33333333333333E-52    x  =  +0.33333333333333E-53
x  =  +0.33333333333333E-54    x  =  +0.33333333333333E-55
x  =  +0.33333333333333E-56    x  =  +0.33333333333333E-57
x  =  +0.33333333333333E-58    x  =  +0.33333333333333E-59
x  =  +0.33333333333333E-60    x  =  +0.33333333333333E-61
x  =  +0.33333333333333E-62    x  =  +0.33333333333333E-63
x  =  +0.33333333333333E+00    x  =  -0.33333333333333E+63
x  =  -0.33333333333333E+62    x  =  -0.33333333333333E+61
```

Figure 1.4: Precision Test: Third Compiler

In the following section we will present a program for studying this problem and test two compilers with the program.

PASCAL PROGRAM: TESTING THE SIN FUNCTION

The program given in Figure 1.5 can be used to check the SIN function of your Pascal. Type up the program and execute it.

Running the SIN Test: Two Compilers

If your built-in SIN function correctly handles small numbers, meaningful values should be returned over the entire dynamic range of the floating-point operations. In this case, floating-point underflow should occur at the same place as for the previous test. In the example of Figure 1.6, the dynamic range of the floating-point operations and the limit of the SIN function are both 3.3×10^{-18}.

```
PROGRAM TSIN(OUTPUT);
(* test range of sin *)

VAR
    I : INTEGER;
    X : REAL;

BEGIN
    X := 1.0E − 4/0.3;
    FOR I := 1 TO 40 DO
        BEGIN
            WRITELN(' x =', X, ', sin =', SIN(X));
            X := 0.1 * X
        END
END.
```

Figure 1.5: Test for the SIN Function

```
x = 3.333333E-04, sin = 3.333329E-04
x = 3.333332E-05, sin = 3.333328E-05
x = 3.333332E-06, sin = 3.333328E-06
x = 3.333332E-07, sin = 3.333328E-07
x = 3.333331E-08, sin = 3.333327E-08
x = 3.333330E-09, sin = 3.333326E-09
x = 3.333330E-10, sin = 3.333326E-10
x = 3.333329E-11, sin = 3.333325E-11
x = 3.333328E-12, sin = 3.333324E-12
x = 3.333328E-13, sin = 3.333324E-13
x = 3.333328E-14, sin = 3.333323E-14
x = 3.333327E-15, sin = 3.333322E-15
x = 3.333327E-16, sin = 3.333322E-16
x = 3.333326E-17, sin = 3.333322E-17
x = 3.333326E-18, sin = 3.333322E-18
Multiply overflow        (error message from compiler)
```

Figure 1.6: SIN Test: First Compiler

Some Pascal implementations of the SIN function initially square the argument without performing a range check. In this case, a floating-point underflow will occur much too soon. In the example of Figure 1.7, the normal dynamic range is 10^{+18} to 10^{-18}. But floating-point underflow occurs during calculations of the SIN when the argument is less than 10^{-9}.

TURBO OK

```
x = .3333335E-03, sin = .3333323E-03
x = .3333334E-04, sin = .3333321E-04
x = .3333334E-05, sin = .3333321E-05
x = .3333333E-06, sin = .3333320E-06
x = .3333331E-07, sin = .3333319E-07
x = .3333331E-08, sin = .3333319E-08
x = .3333330E-09, sin = .3333318E-09
x = .3333329E-10, sin =-.2100363E+07
x = .3333328E-11, sin =-.2100361E+04
...
```

Figure 1.7: SIN Test: Second Compiler

OTHER PASCAL FUNCTIONS

In this section we will look at programs designed to test other functions. We will also learn how to write some trigonometric functions that are not included in standard Pascal compilers, and we will look into the problematic Pascal ARCTAN function.

In some cases, functions can be paired with their inverses. For example, we can take the LOG of the EXP as illustrated in Figure 1.8. Or we can take the ARCTAN of the ratio of SIN/COS. (The TAN function is not included in standard Pascal.)

```
PROGRAM TLOG(OUTPUT);
(* test ln and exp *)

VAR
    I : INTEGER;
    X, Y : REAL;

BEGIN
    X := 1.0E-4/0.3;
    FOR I := 1 TO 20 DO
        BEGIN
            Y := LN(X);
            WRITELN(' x =', X, ', ln =', Y, ', exp(ln) =', EXP(Y));
            X := 0.5 * X
        END
END.
```

Figure 1.8: A Test of the LN and EXP Functions

Standard Pascal compilers include the functions SQRT, SQR, EXP, LN, SIN, COS, and ARCTAN. Unfortunately, the ARCTAN function has not been uniformly implemented for negative arguments. Some compilers return an angle in the second quadrant (90° to 180°) and others return a value in the fourth quadrant (0° to −90°). Furthermore, since there is only a single argument to the ARCTAN, it is not possible to distinguish between angles in the first quadrant and angles in the third quadrant.

The arc tangent function given in Figure 1.9 can properly handle all of

```
FUNCTION ATAN(X, Y : REAL): REAL;
(* arctan in degrees *)
CONST
    PI180 = 57.2957795;
VAR
    A : REAL;
BEGIN (* function atan *)
    IF X = 0.0 THEN
        IF Y = 0.0 THEN ATAN := 0.0
        ELSE ATAN := 90.0
    ELSE (* x <> 0 *)
        IF Y = 0.0 THEN  ATAN := 0.0
        ELSE (* x and y <> 0 *)
            BEGIN
                A := ARCTAN(ABS(Y/X)) * PI180;
                IF X > 0.0 THEN
                    IF Y > 0.0 THEN ATAN := A (* x, y > 0 *)
                    ELSE ATAN := −A (* x > 0, y < 0 *)
                ELSE (* x < 0 *)
                    IF Y > 0.0 THEN ATAN := 180.0 −A (* x < 0, y > 0 *)
                    ELSE ATAN := 180.0 +A (* x, y < 0 *)
            END (* else *)
END (* function atan *);
```

Figure 1.9: An Arc Tangent Function with Two Arguments

these possibilities. There are two arguments, corresponding to the x and y components of the angle. This function calls the regular Pascal ARCTAN function with a positive argument. Then the result is corrected for the proper quadrant. For example, consider an angle that has an x component of − 5 and a y component of − 5. This corresponds to an angle of (180 + 45)° in the third quadrant. However, the tangent of this angle is unity; consequently, the regular arc tangent function will return an angle of 45°. The ATAN function given in Figure 1.9 properly returns a result of 225° for this example. Notice that this arc tangent function gives the angle in degrees rather than in the usual radians.

The arc sine and arc cosine functions are not included in standard Pascal compilers. However, they can be derived from the built-in arc tangent function. The Pascal function shown in Figure 1.10 can be used to calculate the arc sine. It utilizes the arc tangent function given in Figure 1.9. The arc sine function returns an angle in the first quadrant if the argument is positive. The angle is negative and in the fourth quadrant if the argument is negative. The resulting angle is expressed in degrees.

The arc cosine function is given in Figure 1.11. It too requires the arc tangent function given in Figure 1.9. An angle in the first quadrant is returned if the argument is positive. If the argument is negative, however, the resulting angle is in the second quadrant.

```
FUNCTION ARCSIN(X : REAL): REAL;
(* arcsin in degrees *)
(* function ATAN is required *)
(* Feb 1, 81 *)

BEGIN (* function arcsin *)
   IF X = 0.0 THEN ARCSIN := 0.0
   ELSE
      IF X = 1.0 THEN ARCSIN := 90.0
      ELSE
         IF X = −1.0 THEN ARCSIN := −90.0
         ELSE ARCSIN := ATAN( 1.0, X/SQRT(1.0 − SQR(X)))
END (* function arcsin *);
```

Arcsin(x)= ATN (X / SQRT(-X*X+1))

Figure 1.10: The Arc Sine Function

```
FUNCTION ARCCOS(X : REAL): REAL;
(* arccos in degrees *)
(* function ATAN is required *)
(* Feb 1, 81 *)

BEGIN (* function arccos *)
   IF X = 0.0 THEN  ARCCOS := 90.0
   ELSE
      IF X = 1.0 THEN ARCCOS := 0.0
      ELSE
         IF X = -1.0 THEN ARCCOS := 180.0
         ELSE ARCCOS := ATAN( X/SQRT(1.0 - SQR(X)),1.0)
END (* function arccos *);
```

Figure 1.11: The Arc Cosine Function

A program that can be used to test the arc functions is given in Figure 1.12. Function ATAN of Figure 1.9 and function ARCSIN of Figure 1.10 are needed.

In this section, we have seen Pascal implementations of trigono-metric and logarithmic functions including both user-derived func-tions and standard Pascal functions. We will return to the topic of logarithmic and exponential functions at the end of this chapter. First, however, we will look at the Pascal options for handling external files.

EXTERNAL FILES

Figure 1.12 demonstrates the use of the INCLUDE directive and the **EXTERN** statement. Both of these techniques refer to parts of a Pascal program that are located in a separate disk file. Pascal compilers will typically incorporate only one of these two options, although in some cases neither is provided. The technique has several advantages. First, the main source program will be less cluttered and therefore easier to comprehend if large procedures and functions are separated from the main program.

A second advantage of external files is that several different source programs can refer to the same external file. This will reduce the total amount of required disk space since there need not be a separate copy of the procedure in each program. Another advantage is that an exter-nal file can be more easily revised since there is only one copy.

```
PROGRAM TASIN(INPUT, OUTPUT);
(* test arcsin and arctan *)

VAR
   X : REAL;

(* FUNCTION atan(x, y : real): real;
extern; *)
(*$I ATAN.PAS *)

(* FUNCTION arcsin(x : real): real;
extern; *)
(*$I ASIN.PAS *)

BEGIN (* main program *)
   REPEAT
      WRITE(' X:');
      READLN(X);
      WRITELN(' Arcsin of',X:5:2, ' is', ARCSIN(X):7:4)
   UNTIL X = -1.0
END.
```

Figure 1.12: Program to Test the Arc Sine and Arc Tangent Functions

The INCLUDE Directive

The INCLUDE command is a compiler directive with the form:

      (*$I NAME.PAS*)   (UCSD version)

or

      (*$F NAME.PAS*)   (Pascal/M version)

When the compiler encounters the INCLUDE directive, it will go to the named source file and process the statements it finds there. When an end-of-file character is encountered in the external file, the compiler returns to the main source program. The resulting code will be the same as if the INCLUDE file were actually embedded into the main source program.

The INCLUDE directive is embedded in Pascal comment characters, and so it is syntactically ignored by the compiler. Pascal INCLUDE directives cannot generally be nested. That is, an INCLUDE file cannot itself contain an INCLUDE directive to another disk file.

It should be noticed that the INCLUDE directive is only a bookkeeping feature. Since the INCLUDE file is an ASCII source program, it takes up the same amount of disk space as if it were physically part of the main program. In fact the total allocated disk space for the main program and the INCLUDE file may be somewhat greater because there is a minimum block size for disk files. Furthermore, the procedure in the INCLUDE file must be compiled each time the main program is compiled. Consequently, no time is saved. Nevertheless, the INCLUDE feature is ideal when one procedure is used by a number of different programs.

The EXTERN Statement

The **EXTERN** feature, available on some Pascal compilers, is similar to the INCLUDE directive, but it is much more sophisticated. In this case, a procedure can be separately compiled. The external file will allow the declarations **CONST**, **TYPE** and **VAR** at the beginning, so that arrays used as parameters can be defined. For example:

```
CONST
    MAXR = 7;
TYPE
    ARY2 = ARRAY[1..MAXR, 1..MAXC] OF REAL;
VAR
    A, Y : ARY2;
```

Then the regular procedure body will follow:

```
PROCEDURE NAME(A, Y : ARY);
    . . .
END.
```

A period is placed after the last **END** statement to indicate the end of the external file. Some compilers require a semicolon just prior to the period.

The external file is referenced in the main program by giving the regular procedure or function heading. This is then followed by the word **EXTERN**. For example:

> **PROCEDURE** PLOT(X, Y, YCALC : ARY;
>
> N : INTEGER);
>
> **EXTERN**;

EXTERN is not a part of standard Pascal, although it is mentioned in Jensen and Wirth.[1] It is less commonly implemented than the IN-CLUDE directive. However, it is available with UCSD Pascal, Pascal/MT+, and Pascal/Z. Throughout this book, both forms of external reference are given. The **EXTERN** portion, however, is embedded in comment characters. These will have to be removed if the **EXTERN** feature is desired.

Now let us continue our study of logarithmic and exponential functions of Pascal.

A POWER-OF-10 FUNCTION

Occasionally it is necessary to raise a number, say 10, to some power, say x. In BASIC, the expression:

$$Y = 10 \uparrow X$$

will define the variable y as 10 to the power of x. For example, if x equals 0.1, then the value of y becomes 1.2589... In FORTRAN, the corresponding expression is:

$$Y = 10**X$$

Unfortunately, standard Pascal does not include a power operation. But since the EXP function is a standard feature of Pascal, the power of ten can be calculated as:

$$Y := EXP(X*LOG10)$$

where LOG10 is a constant that has been previously defined as the natural logarithm of 10. In the more general case, the x power of z can be obtained from the expression:

$$Y := EXP(X*LN(Z))$$

where LN is the natural log function.

A Taylor Series Expansion to Perform EXP

If the EXP function is not available, then the above expressions cannot be used. In this case, it is possible to develop the desired expression

---

[1]Jensen and Wirth, *Pascal User Manual and Report*, p. 90.

from a Taylor series expansion. The formula is:

$$f(x) = f(a) + (x-a)f'(a) + (x-a)^2 f''(a) / 2! \ldots$$

If the expansion is performed around zero, then the value of a is zero and the terms are:

$$f(x) \ = 10^x \qquad\qquad\qquad f(0) \ = 1$$
$$f'(x) \ = 10^x \ln (10) \qquad\qquad f'(0) \ = \ln (10)$$
$$f''(x) = 10^x [\ln (10)]^2 \qquad\quad f''(0) = [\ln (10)]^2$$

$$\ldots \qquad\qquad\qquad\qquad \ldots$$

The resulting series is:

$$f(x) \ = \ 1 \ + x \ln 10 \ + x^2 \frac{(\ln 10)^2}{2!} \ + x^3 \frac{(\ln 10)^3}{3!} \ + \ldots$$

The series is most accurate at zero and becomes less accurate as the magnitude of the argument increases. The coefficients for a ten-term power series are:

$$t_0 = 1.0$$
$$t_1 = 2.3025851$$
$$t_2 = 2.6509491$$
$$t_3 = 2.0346786$$
$$t_4 = 1.1712551$$
$$t_5 = 0.5393829$$
$$t_6 = 0.2069958$$
$$t_7 = 0.0680894$$
$$t_8 = 0.0195977$$
$$t_9 = 0.0050139$$

The corresponding equation is:

$$10^x = \sum_{i=0}^{9} t_i x^i$$

This series will produce six-figure precision when the argument lies between the values of 0.5 and -0.5. The result will be less accurate when the argument is outside this range.

Range reduction can be used to bring any argument into this range. For example, if the argument is larger than 0.5, then a number such as unity can be repeatedly subtracted from the argument until the result is less than 0.5. Of course, the result must be multiplied by 10 each time a subtraction is performed. For example, suppose that the initial argument is 3.5. Then:

$$10^{3.5} = 10^3 \cdot 10^{0.5} = 10 \cdot 10^{2.5} = 100 \cdot 10^{1.5} = 1000 \cdot 10^{0.5}$$

Thus, the value of 10 to the power 3.5 can be found from the product of 1000 and $10^{0.5}$. In a similar way, arguments that are less than -0.5 can be incremented until they are greater than -0.5. The result is then divided by 10 for each such addition. We will now study the implementation of this Taylor series in a Pascal program.

PASCAL PROGRAM: CALCULATING POWERS OF 10

Function POWER, incorporated into Figure 1.13, can be used to calculate the power of ten from a ten-term Taylor series expansion. In this function, range checking is performed in several stages. The initial argument is inspected at the beginning of the function. If the value is very large, then a large, floating-point number is returned as the value of the function. If the argument is very small, then the value of zero is returned. The actual limits for these two tests will have to be tailored to the floating-point arithmetic of your Pascal compiler. The results from the program given in Figure 1.1 can be used to set this value. For example, if a floating-point underflow occurred immediately after a value of 3.333E-18, then the value of SMALL should be 18. Notice that the function POWER is recursively called when the argument is very large.

Range reduction is employed if the magnitude of the argument is between 0.5 and the value of LARGE. The argument is repeatedly reduced by 4 until the result is less than 4. Then the argument is reduced by unity until the result is below 0.5. Negative arguments are treated similarly. When the argument is in the range 0.5 to -0.5, the Taylor series expansion is utilized. The result is correct to six significant figures.

Notice that the initial argument, supplied by the calling program, is never altered. Rather, a change of variable,

    Y := X;

is made near the beginning of the function. This change is not actually necessary since the parameter is passed by value rather than by reference.

Evaluating the Taylor Series

Any of several possible methods could have been used to evaluate the Taylor series. The simplest approach would be to program the series in its usual mathematical form:

$$SUM := 1.0 + T1*Y + T2*Y*Y + T3*Y*Y*Y + ...$$

The advantage of this form is the clarity of the resulting code. Unfortunately, this is an extremely inefficient method. The execution time increases dramatically as the length of the power series increases.

Another approach is the looping method. In this case, a set of instructions such as:

```
SUM := 1.0;
TERM := 1.0;
FOR I := 1 TO 9 DO
   BEGIN
      TERM := TERM * Y * LOG10/I;
      SUM := SUM + TERM
   END;
```

can be used to add one term of the series for each pass through the loop. This approach is considerably easier to program than the method actually chosen. Furthermore, the length of the power series can be readily changed to accommodate a different floating-point package. However, this method is also rather slow.

The actual method used in our program to evaluate the power series is a direct summing of a fixed number of polynomial terms. The coefficients are defined as constants at the beginning of the function. Then the summation is performed with a nesting of the polynomial terms. This is the fastest method of the three. A disadvantage of this method, however, is the need to incorporate over 50 digits into the function.

The driver program, which is also included in Figure 1.13, will automatically call the power function with a set of positive and negative arguments. The results are printed along with values calculated from the built-in EXP function for comparison. If your Pascal compiler does not incorporate the EXP function, you will have to remove these references in the WRITELN statements.

```
PROGRAM TPOW( OUTPUT);
(* test power-of-ten function *)


CONST
   LOG10 = 2.3025851;


VAR
   I : INTEGER;
   X, Y, A : REAL;


FUNCTION POWER(X : REAL) : REAL;
(* calculate ten to the power x *)
(* by a tenth-order polynomial *)
(* Feb 5.0, 81 *)


CONST
   BIG = 16.0;
   T1 = 2.3025851;
   T2 = 2.6509491;
   T3 = 2.0346786;
   T4 = 1.1712551;
   T5 = 0.5393829;
   T6 = 0.2069958;
   T7 = 0.0680894;
   T8 = 0.0195977;
   T9 = 0.0050139;


VAR
   SUM, Y, ANS : REAL;
```

Figure 1.13: A Function to Calculate Powers of 10

```
BEGIN (* function power *)
   IF X > BIG THEN POWER := POWER(BIG)
   ELSE
      IF X < −BIG THEN POWER := 0.0
      ELSE
         BEGIN
            Y := X;
            ANS := 1;
            WHILE Y > 4.0 DO
               BEGIN (* reduce by 4 *)
                  Y := Y − 4.0;
                  ANS := 10000.0 * ANS
               END;
            WHILE Y < −4.0 DO
               BEGIN (* increase by 4 *)
                  Y := Y + 4.0;
                  ANS := 0.0001 * ANS
               END;
            WHILE Y > 0.5 DO
               BEGIN (* reduce by 1 *)
                  Y := Y − 1.0;
                  ANS := 10.0 * ANS
               END;
            WHILE Y < −0.5 DO
               BEGIN (* increase by 1 *)
                  Y := Y + 1.0;
                  ANS := 0.1 * ANS
               END;
            SUM := (T5 + Y*(T6 + Y*(T7 + Y*(T8 + Y*T9))));
            SUM := 1.0 + Y*(T1 + Y*(T2 + Y*(T3 + Y*(T4 + Y*SUM))));
            POWER := ANS * SUM
         END (* else *)
END (* function power *);
```

Figure 1.13: A Function to Calculate Powers of 10 (cont.)

```
BEGIN (* main program *)
   WRITELN;
   FOR I := 1 TO 4 DO (* small step *)
     BEGIN
        X := 0.2 * I - 0.1;
        A := POWER( X );
        WRITELN(' x=', X:5:2, ', 10^x=', A,
           ', actual=', EXP( X * LOG10));
        Y := -X; (* change sign *)
        A := POWER( Y);
        WRITELN(' x=', Y:5:2, ', 10^x=', A,
           ', actual=', EXP( Y * LOG10))
     END;
   FOR I := 1 TO 6 DO (* large step *)
     BEGIN
        X := 2.0 * I - 0.5;
        A := POWER( X );
        WRITELN(' x=', X:5:2, ', 10^x=', A,
           ', actual=', EXP( X * LOG10));
        Y := -X; (* change sign *)
        A := POWER( Y );
        WRITELN(' x=', Y:5:2, ', 10^x=', A,
           ', actual=', EXP( Y * LOG10))
     END
END.
```

Figure 1.13: A Function to Calculate Powers of 10 (cont.)

The output from the program is shown in Figure 1.14.

```
x= 0.10,  10^x= 1.25893E+00,  actual= 1.25893E+00
x=-0.10,  10^x= 7.94328E-01,  actual= 7.94328E-01
x= 0.30,  10^x= 1.99526E+00,  actual= 1.99526E+00
x=-0.30,  10^x= 5.01187E-01,  actual= 5.01187E-01
x= 0.50,  10^x= 3.16228E+00,  actual= 3.16228E+00
x=-0.50,  10^x= 3.16227E-01,  actual= 3.16228E-01
x= 0.70,  10^x= 5.01187E+00,  actual= 5.01187E+00
x=-0.70,  10^x= 1.99526E-01,  actual= 1.99526E-01
x= 1.50,  10^x= 3.16228E+01,  actual= 3.16228E+01
x=-1.50,  10^x= 3.16227E-02,  actual= 3.16228E-02
x= 3.50,  10^x= 3.16228E+03,  actual= 3.16228E+03
x=-3.50,  10^x= 3.16227E-04,  actual= 3.16228E-04
x= 5.50,  10^x= 3.16228E+05,  actual= 3.16228E+05
x=-5.50,  10^x= 3.16227E-06,  actual= 3.16228E-06
x= 7.50,  10^x= 3.16228E+07,  actual= 3.16228E+07
x=-7.50,  10^x= 3.16227E-08,  actual= 3.16228E-08
x= 9.50,  10^x= 3.16228E+09,  actual= 3.16228E+09
x=-9.50,  10^x= 3.16227E-10,  actual= 3.16228E-10
x= 11.50,  10^x= 3.16228E+11,  actual= 3.16228E+11
x=-11.50,  10^x= 3.16227E-12,  actual= 3.16228E-12
```

Figure 1.14: Output from the Power-of-10 Program

SUMMARY

In this chapter we have written a number of evaluative tools designed for testing Pascal compilers. We began with a program to test floating-point operations; we ran this program on three different compilers and compared the results. We then discovered an interesting quirk of the SIN and COS functions in some Pascal compilers, and we devised a program for testing these functions.

In the midst of studying some other function-testing programs, we found that we needed a technique for incorporating external files into a main source program; this need led us to a discussion of the Pascal INCLUDE directive and **EXTERN** statement.

Finally, we applied a number of the elements of this chapter to implement a Taylor series expansion in a Pascal program for computing powers of 10.

CHAPTER **2**

Mean and Standard Deviation

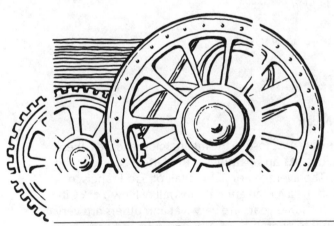

INTRODUCTION

In this chapter we will review some statistical tools and discuss how to implement these tools in Pascal. We will describe the uses of the mean and the standard deviation, and present a program that calculates both of these values. We will then discuss random numbers and some methods of generating them on a computer. In particular, we will look at two Pascal implementations of random number generators. Finally, we will evaluate both of these random functions, using Pascal programs designed to test the "randomness" of the resulting number.

THE MEAN

We often use a single number, called the *average* or *mean value*, to summarize a particular group of data. The mean is calculated by adding all of the items in the group and then dividing by the number of items. The formula is:

$$\bar{y} = \frac{\sum_{i=1}^{N} y_i}{N} \tag{1}$$

where y_i is the set of data containing N elements. The symbol $\bar{y}$ (pronounced y-bar) is the resulting mean.

On the other hand, when there is a uniform *continuous distribution* of the data, the mean can be determined by integration. Consider the function $y = f(x)$ over the interval from limit a to limit b. The mean value of y is constant over this interval. Consequently, the area under the mean will be equal to the area under the curve $f(x)$.

$$\bar{y}\,(b - a) = \int_a^b f(x)dx$$

Therefore, the average value of y is:

$$\bar{y} = \frac{\int_a^b f(x)dx}{b - a}$$

Dispersion from the Mean

Sometimes all of the data are close to the mean. In other cases, there is a great range of values. As an example of the latter, consider the reporting of weather. The average annual rainfall of San Francisco is said to be 19 inches and the average annual snowfall of New York City is given as 30 inches. But some years are very wet and others are very dry. The average annual temperature of Albuquerque and San Francisco is exactly the same: 57°. But the climate of these two cities is very different.

As another example, consider a particular brand of breakfast cereal that contains the statement:

    Net weight 16 ounces

on the box. Suppose that an inspector from the Bureau of Weights and Measures decides to check this brand of cereal in a grocery store. Several boxes are opened and the contents are weighed. Some boxes are found to contain exactly 16 ounces, but others have 15 ounces or

17 ounces. Should the boxes that contain only 15 ounces be confiscated as examples of short weight? If the contents of 100 boxes of cereal are weighed, the resulting frequency distribution might look like the curve in Figure 2.1.

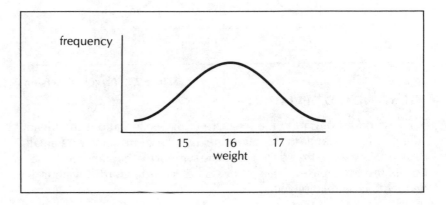

Figure 2.1: A Bell Shaped Curve

The average weight is 16 ounces. But some of the cereal boxes weigh more than the mean value and others weigh less. Furthermore, there are very few boxes that weigh more than 17 ounces or less than 15 ounces.

The frequency distribution shown in Figure 2.1 is *bell shaped*. The curve shows a *Gaussian* or *normal* distribution. This behavior is typical of random variation about a mean value. The equation of this bell shaped curve is related to the Gamma function and the Gaussian error function, which we will study in Chapter 11.

Now, suppose that a second type of breakfast cereal is also tested. The results, this time, might look like the curve in Figure 2.2. The data again show a frequency distribution that is bell shaped with a mean value of 16 ounces. But this time, there is a larger dispersion in weights. Some boxes are as heavy as 26 ounces while others are as light as 6 ounces.

Clearly, there is a difference between the packaging of the first type of cereal and the second, even though they both have the same average weight. Something else besides the mean value is needed to describe the distribution. We need something that describes the *dispersion*. The tool that we can use, the standard deviation, is described in the next section.

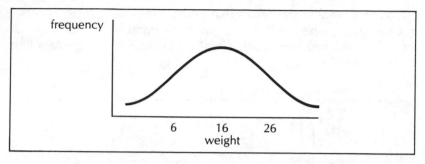

Figure 2.2: Larger Dispersion

THE STANDARD DEVIATION

The *standard deviation* is a measure of dispersion about the mean value. A large standard deviation means a large dispersion; a small deviation means a small dispersion. The symbol for the standard deviation is the lowercase Greek letter sigma. The standard deviation is defined by the relationship:

$$\sigma = \sqrt{\frac{\sum_{i=1}^{N}(\bar{y} - y_i)^2}{N - 1}} \qquad (2)$$

where $\bar{y}$ is the mean and y_i is the set of N data.

Equation 2 demonstrates the meaning of the standard deviation. The square of the difference between each element and the mean value is important. If the elements are closely grouped about the mean, then this difference will be small. The corresponding standard deviation will also be small. On the other hand, if data are spread far from the mean, this difference will be large. The resulting standard deviation will also be large.

The standard deviation can be used quantitatively to describe the dispersion of a set of data. For example, a range of one standard deviation on either side of the mean includes 68% of the sample. About 95% of the values lie within two standard deviations of the mean, and almost all of the values, 99.7%, lie within three standard deviations of the mean.

$$\bar{y} \pm \sigma \qquad 68\%$$

$$\bar{y} \pm 2\sigma \qquad 95\%$$

$$\bar{y} \pm 3\sigma \qquad 99.7\%$$

Suppose that we want to select a type of steel for constructing a bridge. Tensile tests are conducted to determine the strength. One steel is found to have a strength of 450 MPa (megapascals), with a standard deviation of 10 MPa. The results indicate that 99.7% of the pieces are expected to have a strength in the range of 420 to 480 MPa, that is, within three sigmas. Consequently, a design could be based on a strength of 420 MPa, three sigmas below the mean value.

But suppose that a second type of steel was found to have a mean strength of 500 MPa. Is it a better steel than the first? What if the standard deviation for this second steel is 40 MPa? This larger sigma means a larger spread in values. A strength of three sigmas below the mean of this second steel is only 380 MPa. Thus this second steel is not as good, even though its mean value is higher.

Metals used for construction typically have a small sigma. But brittle materials, such as concrete and glass, are very different. They usually have large sigmas. Suppose that tests are made on a type of glass suitable for the front door of an office. The average breaking strength is 120 MPa. But if the sigma is only 40 MPa, then three sigmas below the mean gives a value of zero. Thus it is important to consider the standard deviation as well as the mean value.

In this section we have learned the meaning and importance of the standard deviation. We will now see how to calculate this value and we will design a Pascal program to perform the calculation.

Calculation of the Standard Deviation

While Equation 2 correctly demonstrates the meaning of the standard deviation, it is an inferior method of calculation. The subtraction of each element from the mean value will cause a loss of significance. This is especially important for points that are very close to the mean. That is, the smaller the value of sigma, the greater the problem will be. Another disadvantage of Equation 2 is that a calculation of the average is required before any subtractions can be performed. This calculation cannot be performed until all of the data are available.

A better method of calculating the standard deviation can be derived by expanding the numerator of Equation 2.

$$\Sigma \, (\bar{y}^2 - 2\bar{y}y_i + y_i^2)$$

Now, by combining the above with Equation 1, we have:

$$\left(\frac{\Sigma y_i}{N}\right)^2 N - 2\left(\frac{\Sigma y_i}{N}\right)\Sigma y_i + \Sigma y_i^2$$

since $\bar{y} = \Sigma y_i / N$. The resulting formula is:

$$\sigma = \sqrt{\frac{\Sigma y^2 - \Sigma y \, \Sigma y / N}{N - 1}}$$

N = # of data

With this method, two running totals are kept: the sum of the individual values and the sum of the squares of the values.

PASCAL PROGRAM: MEAN AND STANDARD DEVIATION

The program shown in Figure 2.3 can be used to calculate the mean and standard deviation of a set of numbers. The main program asks the

```pascal
PROGRAM MEANS(INPUT, OUTPUT);
(* Find mean and standard deviation *)
(* Jan 19, 81 *)

CONST
   MAX = 80;

TYPE
   ARY = ARRAY[1..MAX] OF REAL;

VAR
   X            : ARY;
   I,N          : INTEGER;
   MEAN, STD : REAL;

PROCEDURE MEANSTD
               ( X  : ARY; (* array of values *)
        LENGTH         : INTEGER;
        VAR MEAN       : REAL;
        VAR STD _ DEV  : REAL);

VAR
   I                   : INTEGER;
   SUM _ X, SUM _ SQ: REAL;
```

Figure 2.3: Calculation of the Mean and Standard Deviation

```
BEGIN
   SUM_X := 0;
   SUM_SQ := 0;
   FOR I := 1 TO LENGTH DO
      BEGIN
         SUM_X := SUM_X + X[I];
         SUM_SQ := SUM_SQ + X[I] * X[I]
      END;
   MEAN := SUM_X/LENGTH;
   STD_DEV :=
         SQRT((SUM_SQ − SQR( SUM_X)/LENGTH)/(LENGTH−1))
END (* procedure meanstd *);

BEGIN (* main program *)
   WRITELN;
   WRITELN(' Calculation of mean and standard deviation');
   REPEAT
      WRITE(' How many points? ');
      READLN(N)
   UNTIL N <= MAX;
   FOR I := 1 TO N DO
      BEGIN
         WRITE( I:3,': ');
         READLN(X[I])
      END;
   MEANSTD( X, N, MEAN, STD);
   WRITELN;
   WRITELN(' For ', N:3, ' points, mean=', MEAN:8:4,
           ', sigma=', STD:8:4)
END.
```

Figure 2.3: Calculation of the Mean and Standard Deviation (cont.)

user for the number of items. Then procedure MEANSTD is called for the calculation of the mean and standard deviation. The procedure utilizes four parameters. The first parameter is the array of values; it is declared to be of type ARY. The next parameter is the number of values in the array. The third and fourth parameters are the mean and standard deviation calculated by the procedure.

The program shown in Figure 2.3 demonstrates how arrays can be included as parameters to procedures. Pascal requires a specific declaration of each parameter. However, the expression:

PROCEDURE MEANSTD(X : **ARRAY**[1..MAX] **OF** REAL; . . .);

is not acceptable. Consequently, each array used as a parameter must be specifically typed in the calling program.

TYPE

ARY = **ARRAY**[1..MAX] **OF** REAL;

Then the defined type is included in the heading of the procedure.

PROCEDURE MEANSTD(X : ARY; . . .);

The program begins by asking the user for the number of values to be entered. If the answer exceeds the declared length of the array X, then the question is asked again. After the values have been entered, the mean and standard deviation are determined.

If many values are to be entered, an alternate method might be considered. In this case, the computer program could count the number of values as they are entered. A special value, such as a number less than −20000, could be used to signal the end of the list.

Compile the program and run it. Input the five numbers 1,2,3,4,5, and verify that the mean is 3 and the standard deviation is 1.58. The results should look like Figure 2.4.

```
Calculation of mean and standard deviation
How many points? 5
  1:  1
  2:  2
  3:  3
  4:  4
  5:  5

For    5 points, mean=  3.0000, sigma=  1.5811
```

Figure 2.4: Output from the Mean and Standard Deviation Program

This completes our introduction to the mean and the standard deviation. We will later return to these subjects; but we will now proceed to the subject of random numbers and random number generators.

RANDOM NUMBERS

A set of random numbers can sometimes be used to test a computer program if actual data are not available. For example, suppose that we have written a program to fit a straight line through a set of experimental data. We could generate a set of points along a straight line. Then, we could selectively move the points off the line by using a random number generator.

PASCAL FUNCTION: A RANDOM NUMBER GENERATOR

Random number generators are provided with some Pascal packages but not with others. A random number generator will be required for some of the programs given in this book. If your Pascal does not incorporate a random number generator, then you can use the program shown in Figure 2.5 for this purpose. The form has been chosen to be compatible with commonly implemented Pascal random number generators.

```
FUNCTION RANDOM(DUMMY: INTEGER): REAL;
(* random number 0 — 1 *)
(* define SEED = 4.0 as global *)
(* Adapted from HP—35 Applications Programs *)

CONST
    PI = 3.14159;
VAR
    X: REAL;
    I: INTEGER;

BEGIN (* Random *)
    X := SEED + PI;
    X := EXP(5.0 * LN(X));
    SEED := X — TRUNC(X);
    RANDOM := SEED
END(* Random *);
```

Figure 2.5: A Pascal Function for Generating Random Numbers

This random number generator will return a sequence of numbers in the range of zero to unity. A typical call to function RANDOM would be:

X[I] := RANDOM(0);

Some random number generators use the parameter as a seed for the next random number. In other cases, a particular value of the argument indicates that the previously returned number should be returned again. But in this case, the parameter for function RANDOM is a dummy argument. Therefore, any integer can be used by the calling program.

The actual seed for sequencing the numbers is saved in the real, global variable named SEED. SEED must be declared and initialized by the calling program. An initial value of 4.0 is a good choice. Procedure RANDOM violates a general rule of programming that states that a subroutine should not change global variables. But in this case, it is probably the best course of action. The problem is that Pascal does not provide for permanent local variables in subroutines.

After the first call to function RANDOM, the value of SEED is the previously returned random number. However, this number cannot be stored as a local variable within RANDOM. The problem lies in the Pascal language. Local variables are always re-initialized upon each entry to a procedure. Another possibility is to declare the seed as a parameter to the random number generator. In this case, the calling program has to keep a copy of the most recent random number to use as the seed for the next call. But this is contrary to the style of random number generators that are often incorporated in Pascal packages. Consequently, the choice was made to store the seed as a global variable.

Now that we have learned how our random number generator works, let us look at a program that tests the quality of the random function.

PASCAL PROGRAM:
EVALUATION OF A RANDOM NUMBER GENERATOR

Whether you use the random number generator given in Figure 2.5 or the one that is supplied with your Pascal package, you should perform a simple test to see how reasonable the results are. The mean value of a set of random numbers ranging from zero to unity should, of course, be one-half. Furthermore, the standard deviation should be the reciprocal of the square root of 12, a value of 0.2887.

The program shown in Figure 2.7 can be used to test a random number generator. It calls function RANDOM 48 times. Then it calls procedure MEANSTD, given in Figure 2.3, for the calculation of the mean and standard deviation. This process is repeated 20 times. The mean value and the corresponding standard deviation are printed for each cycle. The expected values are printed at the head of the columns.

Running the Program

Type up the program and incorporate procedure MEANSTD. Function RANDOM should also be included if necessary. Now test your random number generator. Random number generators are sometimes called pseudo-random number generators to emphasize the fact that they are not truly random. Therefore, you should not expect to obtain a mean of one-half and a standard deviation of 0.2887 for each grouping of 48 values. The first part of the output from program RANTST might look like Figure 2.6.

mean (0.5)	std dev (0.2887)
0.4294	0.3156
0.6200	0.3016
0.5008	0.2439
0.3615	0.2758
0.4121	0.2522
0.3960	0.2554
0.5444	0.2363
0.5013	0.3129
0.5282	0.3734
0.5105	0.2375
0.5585	0.2701

Figure 2.6: Output from Program RANTST

Notice that mean value for the groups ranges from 0.36 to 0.62, and the standard deviation has values from 0.24 to 0.37. Built-in random number generators are usually better than this, but some are worse.

Using π to Produce Random Numbers

A short sequence of random numbers can be obtained from the first 15 digits of π:

3.14159265358979

```
PROGRAM RANTST(OUTPUT);
(* test random number generator RANDOM *)
(* procedure MEANSTD and function RANDOM
         are also required *)

TYPE
   ARY = ARRAY[1..100] OF REAL;

VAR
   X        : ARY;
   N, I, J   : INTEGER;
   R, MEAN, STD,
      SEED  : REAL;

(* FUNCTION random(dummy: integer): real;
extern; *)

(*$F RANDOM.PAS *)

(* PROCEDURE meanstd
               ( x  : ary;
      length     : integer;
   VAR mean       : real;
   VAR std_dev   : real);
extern; *)

(*$F MEANSTD.PAS *)
```

Figure 2.7: Program to Test the Random Number Generator

```
BEGIN  (* Main program *)
   SEED := 4.0;
   N := 48;
   WRITELN;
   WRITELN('  mean   std dev');
   WRITELN('  (0.5)   (0.2887)' );
   WRITELN('  ====================');
   FOR J := 1 TO 20 DO
      BEGIN
         FOR I := 1 TO N DO
            X[I] := RANDOM(0);
         MEANSTD( X, N, MEAN, STD);
         WRITELN( MEAN :10:4, STD :10:4)
      END (* j loop *)
END.
```

Figure 2.7: Program to Test the Random Number Generator (cont.)

The following mnemonic makes it easy to remember the sequence. The number of letters in each word is equal to the corresponding digit of π:

YES, I NEED A DRINK, ALCOHOLIC, OF COURSE, AFTER
3 . 1   4   1   5   9   2   6   5

THE HEAVY SESSIONS INVOLVING QUANTUM MECHANICS
3   5   8   9   7   9

This sequence has a mean value of 5.1 and a standard deviation of 2.8.

Our second random number generator, described in the next section, is designed to simulate experimental data.

Gaussian Random Numbers

Sampling errors, that is, errors made during measurement, can be of any size. However, small errors are more likely than large errors. For example, suppose that a table with an actual length of six feet is measured with a ruler. An error of one foot is less likely than an error of one inch. An error of ten feet is even less likely. Thus, measured values that are further from the correct value are less likely than values that are closer. Consequently, if we want to simulate experimental data with a random number generator, the numbers should not be uniformly spaced. But in fact, the usual random number generator produces a uniform set of numbers over the interval 0 to 1.

What is needed for the simulation of experimental data is a set of random numbers that are grouped about the mean. Thus, the frequency distribution should be bell shaped; it should have a normal or Gaussian distribution. Fortunately, it is fairly easy to produce a Gaussian distribution of random numbers from an ordinary random number generator.

Consider a sequence of 12 random numbers. It is highly unlikely that all 12 will have a value of zero. Similarly, it is very unlikely that they will all be unity. Suppose that we generate a new number from the sum of 12 uniformly distributed random numbers. Since the average value of the original numbers is one-half, the sum of 12 random numbers is likely to be about 12 times one-half, or 6. As we will see, this is the key to generating Gaussian random numbers from an ordinary random number generator. We will now look at a Pascal implementation of such a function.

PASCAL FUNCTION: GAUSSIAN RANDOM NUMBER GENERATOR

A Pascal function for producing Gaussian distributed random numbers is given in Figure 2.8. Each Gaussian random number is obtained by summing 12 uniformly distributed random numbers and subtracting the value of 6. A set of such numbers should have a mean value of zero and a standard deviation of unity. Other values for the mean and standard deviation can readily be chosen. The formula for calculating each random number, RANDG, is:

$$RANDG := SIGMA * (SUM - 6) + MEAN$$

In this expression, SUM is the sum of 12 uniformly distributed random numbers, SIGMA is the desired standard deviation, and MEAN is the desired mean. If each Gaussian random number is formed from more than 12 numbers, then there is an additional complication. In this case, the formula becomes:

$$RANDG := SIGMA * (SUM - NUM/2) * SQRT(12/NUM) \\ + MEAN$$

where NUM is the number of uniformly distributed random numbers used to obtain each Gaussian random number.

Function RANDG can thus be used in conjunction with function RANDOM to generate a series of random numbers with a Gaussian distribution. There are two parameters that must be supplied by the calling program: the desired mean and the standard deviation of the resulting numbers. An alternate approach would be to select fixed values for MEAN and SIGMA. These could then be encoded into function RANDG and would not then appear as parameters.

```
FUNCTION RANDG(MEAN, SIGMA : REAL): REAL;
(* produce random numbers with a Gaussian distribution *)
(* MEAN and SIGMA are supplied by calling program *)
(* function RANDOM is required *)

VAR
    I      : INTEGER;
    SUM  : REAL;

BEGIN
    SUM := 0.0;
    FOR I   := 1 TO 12 DO
        SUM := SUM + RANDOM(0);
    RANDG := (SUM − 6.0) * SIGMA + MEAN
END (* function randg *);
```

Figure 2.8: A Function to Generate Normally Distributed Random Numbers

PASCAL PROGRAM: EVALUATING *RANDG*

A program for testing function RANDG is given in Figure 2.9. The mean for the Gaussian numbers is chosen to be ten and the standard deviation is chosen to be one-half. Procedure MEANSTD is called for the determination of the actual mean and standard deviation for a set of the new random numbers. The resulting numbers should exhibit a mean of 10 and a standard deviation of one-half.

The area under the normal distribution curve is related to the standard deviation. It can be obtained from the Gaussian error function that is developed in Chapter 11.

```
PROGRAM RANTST(OUTPUT);
(* test Gaussian random number generator RANDG *)
(* procedure MEANSTD and function RANDOM
         are also required *)

TYPE
   ARY = ARRAY[1..100] OF REAL;

VAR
   X         : ARY;
   N, I, J   : INTEGER;
   R, AVER, STD,
      SEED  : REAL;


(* FUNCTION random(dummy: integer): real;
extern; *)
(*$F RANDOM.PAS *)


(* PROCEDURE meanstd
                 ( x  : ary;
      length       : integer;
      VAR mean      : real;
      VAR std_dev   : real);
extern; *)
(*$F MEANSTD.PAS *)


(* FUNCTION randg(mean, sigma : real): real;
extern; *)
(*$F RANDG.PAS *)
```

Figure 2.9: Program to Test Gaussian Random Number Generator

```
BEGIN   (* Main program *)
    SEED : = 4.0;
    N : = 50;
    WRITELN;
    WRITELN('   mean    std dev');
    WRITELN('    (10)        (0.5)' );
    WRITELN('   ====================');
    FOR J : = 1 TO 20 DO
      BEGIN
        FOR I : = 1 TO N DO
            X[I] : = RANDG(10, 0.5);
        MEANSTD( X, N, AVER, STD);
        WRITELN( AVER :10:4, STD :10:4)
      END (* j loop *)
END.
```

Figure 2.9: Program to Test Gaussian Random Number Generator (cont.)

SUMMARY

We began this chapter with a discussion of two important statistical tools, the mean and the standard deviation. Our discussion led us to an efficient Pascal program for calculating these values. Then we discovered how to write two different Pascal random number generators, which will prove to be useful tools if our system does not supply such a function. Finally, we discussed the importance of evaluating the reasonableness of the random numbers generated, and we saw two programs that will carry out that evaluation.

In the next chapter we will continue to expand our understanding of Pascal; we will see how to express vectors and matrices in the form of Pascal arrays.

CHAPTER **3**

Vector and Matrix Operations

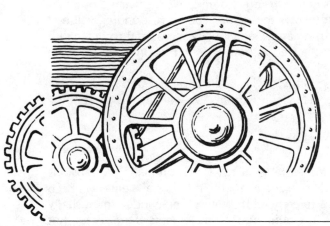

INTRODUCTION

In most of the chapters in this book we develop programs that utilize vectors and matrices. Consequently, in this chapter, we will review the concepts of vectors and matrices and consider some of their more common mathematical operations. We will demonstrate several Pascal implementations. Vectors and matrices are important concepts because they greatly simplify the programming of mathematical operations on data sets.

SCALARS AND ARRAYS

We will begin our discussion by establishing the difference between a *scalar* variable and an *array*. An ordinary, simple variable is called a scalar. It is referenced by a unique symbolic name, and it is associated with a single value. For example, the Pascal expression:

YEAR := 1971

assigns a value to the scalar variable YEAR. In Pascal, scalars can be declared as real, integer, character or logical (boolean) variables.

Sometimes it is necessary to refer collectively to a set of scalar values. Pascal incorporates several structured data types for this purpose, but the simplest of these is known as the *array*. In an array, all of the components have the same type. That is, all of the elements are real numbers, integers, or logical variables. Pascal allows structures within structures. Consequently, the elements of an array can themselves be arrays. In the more usual case, however, the elements of an array are scalars.

The components of an array are collectively referenced by a symbolic name. Each position of the array is uniquely designated by an index or subscript that follows the name. The corresponding value at each position in the array can be individually accessed through this index. The value can be changed without affecting the other values of the array.

In the following sections we will see how *vectors* and *matrices* are represented as arrays in Pascal programs. We will begin by describing vectors.

VECTORS

A vector is a one-dimensional array. (The number of dimensions refers to the number of subscripts, not the number of elements.) Each element of a vector is referenced through a single index that usually runs from 1 up to the maximum number of elements of the vector. But in Pascal, the index can begin at any point. Thus a Pascal array might begin with an index of zero or it might begin with an index of 101.

In ordinary usage, the elements of a vector are separated by spaces. For example, consider the vector **v** that contains the values:

2 5 1 9 4 3

The Pascal statement:

VAR V : **ARRAY**[1..6] **OF** INTEGER;

can be used to define the symbol V as a vector of integers. The maximum number of elements (the length) is declared to be six. Values can be assigned to this vector by Pascal statements such as:

V[1] := 2;
V[2] := 5;
V[3] := 1;
V[4] := 9;
V[5] := 4;
V[6] := 3;

The first element of this vector is located at position V[1], the second element is at V[2] and the last element is at V[6]. The first element of this vector has a value of 2, the second element has a value of 5 and the last element has a value of 3.

Since vectors have only a single dimension, it should not matter whether they are written horizontally or vertically. But sometimes we will need to distinguish between *row vectors*, which are written horizontally, and *column vectors*, which are written vertically. In this case, the set:

$$\begin{bmatrix} 2 \\ 5 \\ 1 \\ 9 \\ 4 \\ 3 \end{bmatrix}$$

is a column vector.

In the next section we will study the operations of vector arithmetic and their implementation in Pascal.

Vector Arithmetic

The major arithmetic operations defined for vectors are: magnitude, scalar multiplication, vector addition, dot product, and cross product. Let us now look at each of these operations.

Magnitude

The *magnitude* of a vector is a scalar value. It is obtained by summing the squares of the elements, then taking the square root of the resulting sum. For example, the magnitude of the vector **y**:

$$\mathbf{y} = \begin{bmatrix} 2 & 2 & 1 \end{bmatrix}$$

is equal to the square root of 4 + 4 + 1. The resulting value is 3. The operation can be programmed with the Pascal statement:

MAG := SQRT(SQR(Y[1]) + SQR(Y[2]) + SQR(Y[3]))

Scalar Multiplication of Vectors

If a vector **y** is multiplied by a scalar value *s*, then each element of **y** is multiplied by *s*. For example, 2 times the vector **v** is:

$$2\mathbf{v} = \begin{bmatrix} 4 & 10 & 2 & 18 & 8 & 6 \end{bmatrix}$$

The following Pascal expression generates a vector V2 in which each element is twice as large as the corresponding element of vector V:

FOR I := 1 **TO** 6 **DO** V2[I] := 2.0 * V[I]

Vector Addition

Two vectors can be added together if each has the same number of elements, or the same length. The result is a new vector in which each element is the sum of the two corresponding elements of the original vectors. Thus if:

$$\mathbf{a} = \begin{bmatrix} 1 & 2 & 3 \end{bmatrix}$$

and

$$\mathbf{b} = \begin{bmatrix} 3 & 4 & 5 \end{bmatrix}$$

then

$$\mathbf{a} + \mathbf{b} = \begin{bmatrix} 4 & 6 & 8 \end{bmatrix}$$

The corresponding Pascal expression is:

FOR I := 1 **TO** 3 **DO** C[I] := A[I] + B[I]

Dot Product (or Scalar Product)

The *dot product* or *scalar product* of two equal-length vectors produces a scalar result. Each element of one vector is multiplied by the corresponding element of the other. The resulting products are then added together. The mathematical symbol for the dot operator is simply a dot placed between the operands. Thus:

$$\mathbf{a} \cdot \mathbf{b} = (1)(3) + (2)(4) + (3)(5) = 26$$

The dot product is equal to the product of the magnitude of the original vectors and the cosine of the angle between them:

$$\mathbf{a} \cdot \mathbf{b} = |\mathbf{a}| \, |\mathbf{b}| \cos \theta$$

This formula can be used to find the angle between two vectors. For example, the angle between the vectors **a** and **b** is 10.7°:

$$\cos \theta = 26/(3.742 \cdot 7.071) = 0.9826 \quad \text{and}$$
$$\text{Arccos } 0.9826 = 10.7°$$

Cross Product (or Vector Product)

The *cross product*, or *vector product*, of two vectors produces a third vector that is mutually perpendicular to the two original vectors. The mathematical symbol for this operation is ×. The magnitude of the resulting vector is equal to the product of the magnitudes of the original vectors and the sine of the angle between them:

$$|\mathbf{a} \times \mathbf{b}| = |\mathbf{a}| \, |\mathbf{b}| \sin \theta$$

The cross product, $\mathbf{c} = \mathbf{a} \times \mathbf{b}$, can be calculated from the Pascal expressions:

```
C[1] := A[2]*B[3] − B[2]*A[3];
C[2] := −A[1]*B[3] + B[1]*A[3];
C[3] := A[1]*B[2] − B[1]*A[2]
```

The cross product of the vectors **a** and **b** is the vector [− 2 4 − 2]; its magnitude is 4.9.

We have now seen Pascal implementations for the main arithmetic operations on vectors. Later in this chapter we will see the matrix equivalents of these operations. However, before moving on to matrices we should define one special kind of vector implementation —the *string*.

Strings

Strings of alphabetic and numeric (alphanumeric) characters are useful for representing such things as names and addresses. These can easily be handled in Pascal by defining a vector of strings. That is, each name or address is one element of the vector. In standard Pascal, it is also necessary to further define a string as an array of characters.

```
TYPE
    STRING = PACKED ARRAY [1..15] OF CHAR;
VAR
    ADDRESS : ARRAY [1..100] OF STRING;
```

The **VAR** declaration allows the symbolic name of ADDRESS to contain up to 100 actual addresses. With the standard implementation of Pascal, however, each address must contain exactly the number of declared characters (15 in this case). The length must be filled out with blanks if necessary. For example:

```
ADDRESS[1] := '1234 Oak St.   ';
ADDRESS[2] := '2222 First St.   ';
```

On the other hand, if the dynamic string type is implemented in your Pascal, then the **TYPE** declaration is not needed. Furthermore, each string entry can be any length up to the maximum implemented length (usually 80 characters). For example:

```
ADDRESS[1] := '1234 Oak St.';
ADDRESS[2] := '2222 First St.';
```

In either case, sorting and other operations can be performed on such strings:

```
IF ADDRESS[I] < ADDRESS[J] THEN . . .
```

Now that we have described vectors and strings, let us move on to two-dimensional arrays and their Pascal representations.

MATRICES

A two-dimensional array is called a *matrix*. The elements of this set are arranged into a rectangle or square. The elements of the matrix can be considered as a set of horizontal lines called row vectors or as a set of vertical lines called column vectors. Thus a matrix can be described as a one-dimensional set of vectors.

Each element of a matrix is uniquely defined by a pair of indices: the *row index* and the *column index*. For example, consider the matrix:

$$
\begin{bmatrix}
x_{11} & x_{12} & x_{13} & \cdots & x_{1m} \\
x_{21} & x_{22} & x_{23} & \cdots & x_{2m} \\
x_{31} & x_{32} & x_{33} & \cdots & x_{3m} \\
\cdots & \cdots & \cdots & \cdots & \cdots \\
x_{n1} & x_{n2} & x_{n3} & \cdots & x_{nm}
\end{bmatrix}
$$

which contains n rows and m columns. The row index is always given first. It is then followed by the column index.

A matrix is referenced by its name, which can be a single alphabetic character, or a string of characters. The indices are given as subscripts except in computer programs, where subscripting is not possible. In COBOL, FORTRAN, and BASIC programs, the array subscripts are enclosed in parentheses. Square brackets are used for this purpose in Pascal and APL. Thus, the appearance of the expression X[2,3] in a Pascal program is a reference to the element located in the second row and the third column of the two-dimensional array named X.

A matrix that has the same number of rows as columns is called a *square matrix*. The *principal*, or *main diagonal* of a square matrix contains the elements x_{11}, x_{22}, x_{33}, . . . x_{nn}. The principal diagonal is sometimes referred to as simply the *diagonal*. A square matrix that contains the value of unity at each position of the main diagonal, and is zero everywhere else, is known as a *unit matrix*. For example:

$$
\begin{bmatrix}
1 & 0 & 0 & 0 \\
0 & 1 & 0 & 0 \\
0 & 0 & 1 & 0 \\
0 & 0 & 0 & 1
\end{bmatrix}
$$

is a 4-by-4 unit matrix.

Now that we have defined the essential vocabulary of matrices, we can go on to study the major arithmetic operations for matrices, and the Pascal implementations of these operations.

Matrix Arithmetic

We will begin by defining the transpose operation; then we will describe scalar multiplication, and matrix addition, subtraction, and multiplication.

The Transpose Operation

A matrix is *transposed* by interchanging the rows and the columns. Each original element x_{ij} becomes the new element x_{ji} of the transposed matrix. The *transpose* of matrix X is designated as X^T. Thus, if

$$X = \begin{bmatrix} 1 & 2 & 3 \\ 4 & 5 & 6 \\ 7 & 8 & 9 \end{bmatrix}$$

then,

$$X^T = \begin{bmatrix} 1 & 4 & 7 \\ 2 & 5 & 8 \\ 3 & 6 & 9 \end{bmatrix}$$

Notice that the transpose of a square matrix can be obtained by rotation of the matrix about the principal diagonal.

Two matrices are *equal* if every element of one is equal to the corresponding element of the other. Thus if $X = Y$, then for each element:

$$x_{ij} = y_{ij}$$

A square matrix is symmetric if it is equal to its transpose. In this case, each element x_{ij} equals the corresponding element x_{ji}.

Scalar Multiplication of Matrices

If a matrix X is multiplied by a scalar value s, then each element of the matrix is multiplied by the value s. The following Pascal statement will produce a matrix Y from the product of matrix X and the scalar S.

```
FOR I := 1 TO N DO
    FOR J := 1 TO M DO
        Y[I,J] := X[I,J] * S
```

Matrix Addition and Subtraction

One matrix may be added to or subtracted from another matrix if both have the same number of columns and the same number of rows.

The addition of the matrix X to the matrix Y to produce matrix Z is written as:

$$Z = X + Y$$

Thus, if

$$X = \begin{bmatrix} x_{11} & x_{12} & x_{13} \\ x_{21} & x_{22} & x_{23} \end{bmatrix}$$

and

$$Y = \begin{bmatrix} y_{11} & y_{12} & y_{13} \\ y_{21} & y_{22} & y_{23} \end{bmatrix}$$

then

$$Z = \begin{bmatrix} x_{11} + y_{11} & x_{12} + y_{12} & x_{13} + y_{13} \\ x_{21} + y_{21} & x_{22} + y_{22} & x_{23} + y_{23} \end{bmatrix}$$

The corresponding Pascal statement is:

FOR I := 1 **TO** N **DO**
　　FOR J := 1 **TO** M **DO**
　　　　Z[I,J] := X[I,J] + Y[I,J]

In this statement, each element of Z is formed from the sum from the corresponding elements of X and Y. In a similar way, subtraction of one matrix from another:

$$Z = X - Y$$

is performed by subtracting each element of the second matrix from the corresponding element of the first.

Matrix Multiplication

One matrix may be multiplied by another if the number of columns of the first matrix equals the number of rows of the second. The two matrices are said to be *conformable* in this case. Thus, if X is a matrix that contains m rows and n columns, and Y is a matrix with n rows and p columns, then the product:

$$Z = X Y$$

produces a matrix Z with m rows and p columns. That is, the resulting matrix has the same number of rows as matrix X and the same number of columns as matrix Y.

In matrix multiplication, each element of Z is formed from a sum of products. The elements from one row of matrix X are each multiplied by the corresponding elements from a column of matrix Y, then summed up. For matrix X (which might be the transpose of the previous X) and matrix Y

$$X = \begin{bmatrix} x_{11} & x_{12} \\ x_{21} & x_{22} \\ x_{31} & x_{32} \end{bmatrix}$$

and

$$Y = \begin{bmatrix} y_{11} & y_{12} & y_{13} \\ y_{21} & y_{22} & y_{23} \end{bmatrix}$$

the operation $Z = X\,Y$ is formed as

$$Z = \begin{bmatrix} x_{11}y_{11} + x_{12}y_{21} & x_{11}y_{12} + x_{12}y_{22} & x_{11}y_{13} + x_{12}y_{23} \\ x_{21}y_{11} + x_{22}y_{21} & x_{21}y_{12} + x_{22}y_{22} & x_{21}y_{13} + x_{22}y_{23} \\ x_{31}y_{11} + x_{32}y_{21} & x_{31}y_{12} + x_{32}y_{22} & x_{31}y_{13} + x_{32}y_{23} \end{bmatrix}$$

Each jk element of Z is formed from row j of matrix X and column k of matrix Y according to the scheme

$$z_{jk} = x_{j1}y_{1k} + x_{j2}y_{2k} + \ldots + x_{jn}y_{nk}$$

Thus, if

$$X = \begin{bmatrix} 1 & 4 \\ 2 & 5 \\ 3 & 6 \end{bmatrix}$$

and

$$Y = \begin{bmatrix} 7 & 8 & 9 \\ 10 & 11 & 12 \end{bmatrix}$$

then the product

$$Z = XY = \begin{bmatrix} 47 & 52 & 57 \\ 64 & 71 & 78 \\ 81 & 90 & 99 \end{bmatrix}$$

is a square matrix with dimensions of 3 by 3. The first element, z_{11}, for example, is calculated as:

$$(1)(7) + (4)(10) = 47$$

Matrix multiplication is not commutative; that is, the product $Y X$ will not, in general, be equal to the product $X Y$. Reversing the order of the previous example produces a 2-by-2 matrix, rather than a 3-by-3 matrix:

$$YX = \begin{bmatrix} 50 & 122 \\ 68 & 167 \end{bmatrix}$$

Notice that the dot product of two vectors follows the rules for matrix multiplication if the first vector is a column vector (that is, has one column), and the second is a row vector (that is, has one row).

We have seen how Pascal handles vector and matrix arithmetic through the use of one- and two-dimensional arrays. We are now ready to write a program using a number of the Pascal statements we have learned.

PASCAL PROGRAM: MATRIX MULTIPLICATION

In Chapter 4, we will need a routine for matrix multiplication. Specifically, we will need to multiply the transpose of a matrix X by the original matrix X. We will also need to multiply the vector **y** by the matrix X. Therefore, we will program such a routine now.

The program shown in Figure 3.1 contains procedure SQUARE that will perform both of the needed operations. The main program will generate the X matrix and the **y** vector. Procedure SQUARE will calculate the needed matrix A and vector **g** according to the equations:

$$X^T X = A \quad \text{and} \quad yX = \mathbf{g}$$

```
PROGRAM MATR1( OUTPUT);
(* Feb 8,81 *)
(* Pascal program to perform *)
(* matrix multiplication *)

CONST
    RMAX = 9;
    CMAX = 3;

TYPE
    ARY   = ARRAY[1..RMAX] OF REAL;
    ARYS  = ARRAY[1..CMAX] OF REAL;
    ARY2  = ARRAY[1..RMAX, 1..CMAX] OF REAL;
    ARY2S = ARRAY[1..CMAX, 1..CMAX] OF REAL;

VAR
    Y : ARY;
    G : ARYS;
    X : ARY2;
    A : ARY2S;
    NROW, NCOL : INTEGER;

PROCEDURE GET _ DATA(VAR X : ARY2;
                     VAR Y : ARY;
              VAR NROW, NCOL : INTEGER);


(* get values for nrow, ncol, and arrays x, y *)


VAR  I, J : INTEGER;

BEGIN
    NROW := 5;
    NCOL := 3;
```

Figure 3.1: Matrix Multiplication Program ($A = X^T X$, $G = Y X$)

```
    FOR I := 1 TO NROW DO
      BEGIN
        X[I,1] := 1;
        FOR J := 2 TO NCOL DO
          X[I,J] := I * X[I,J−1];
        Y[I] := 2 * I
      END
END (* procedure get_data *);


PROCEDURE WRITE_DATA;

(* print out the answers *)

VAR
    I, J : INTEGER;

BEGIN
    WRITELN;
    WRITELN('          X                    Y');
    FOR I := 1 TO NROW DO
      BEGIN
        FOR J := 1 TO NCOL DO
          WRITE( X[I,J]:7:1, ' ');
        WRITELN(' : ', Y[I] :7:1)
      END;
    WRITELN('          A                    G');
    FOR I := 1 TO NCOL DO
      BEGIN
        FOR J := 1 TO NCOL DO
          WRITE( A[I,J]:7:1, ' ');
        WRITELN(' : ', G[I] :7:1)
      END
END (* write_data *);
```

Figure 3.1: Matrix Multiplication Program (A = X^T X, G = Y X) (cont.)

```
PROCEDURE SQUARE(X : ARY2;
                 Y : ARY;
             VAR A : ARY2S;
             VAR G : ARYS;
         NROW,NCOL : INTEGER);

(* matrix multiplication routine *)
(* A = transpose X times X *)
(* G = Y times X *)

VAR
   I, K, L : INTEGER;

BEGIN (* square *)
   FOR K := 1 TO NCOL DO
      BEGIN
        FOR L := 1 TO K DO
           BEGIN
             A[K,L] := 0;
             FOR I := 1 TO NROW DO
                BEGIN
                  A[K,L] := A[K,L] + X[I,L] * X[I,K];
                  IF K <> L THEN A[L,K] := A[K,L]
                END
           END (* L loop *);
        G[K] := 0;
        FOR I := 1 TO NROW DO
           G[K] := G[K] + Y[I] * X[I,K]
      END (* k loop *)
END (* square *);

BEGIN (* main program *)
   GET_DATA (X, Y, NROW, NCOL);
   SQUARE(X, Y, A, G, NROW, NCOL);
   WRITE_DATA
END.
```

Figure 3.1: Matrix Multiplication Program (A = X^T X, G = Y X) (cont.)

The matrix X, in this case, contains 5 rows and 3 columns. Vector y has a length of 5.

Since the resulting matrix A is symmetric, the calculations can be simplified. For example, there will be terms like:

$$A[1,3] := A[3,1]$$

Matrix A will have 3 rows (the same as the transpose of matrix X) and 3 columns (the same as matrix X). Vector g will have a length of 3. Actually, both **y** and **g** must be considered as row vectors, that is, they have a dimension of 1 row and 5 columns. Alternately, if we want to consider the vectors **y** and **g** as column vectors, then we should write the multiplication equation as:

$$y^T X = g^T$$

where the transposed column vectors become row vectors.

Type up the program shown in Figure 3.1 and execute it. The results should look like Figure 3.2.

```
              X                          Y
       1.0       1.0       1.0     :      2.0
       1.0       2.0       4.0     :      4.0
       1.0       3.0       9.0     :      6.0
       1.0       4.0      16.0     :      8.0
       1.0       5.0      25.0     :     10.0
              A                          G
       5.0      15.0      55.0     :     30.0
      15.0      55.0     225.0     :    110.0
      55.0     225.0     979.0     :    450.0
```

Figure 3.2: Output from the Matrix Multiplication Program

DETERMINANTS

The *determinant* of a square matrix X is designated as |X|. The result is a scalar value. For a 2-by-2 matrix, the upper-left member is multiplied by the lower-right, and then the product of the lower-left member and the upper-right member is subtracted:

$$|X| = x_{11}x_{22} - x_{12}x_{21}$$

For example, the determinant of the matrix:

$$\begin{bmatrix} 1 & 2 \\ 3 & 4 \end{bmatrix}$$

is -2.

The determinant of matrices larger than 2 by 2 can be found by multiplying each element of the first row by the determinant of the remaining matrix, after removing the column that is common to the element in the first row. A recursive definition is:

$$|X| = x_{11}s_{11} - x_{12}s_{12} + x_{13}s_{13} - \ldots (-1)^{n+1}x_{1n}s_{1n}$$

where x_{11}, x_{12}, etc., are the elements of the first row of matrix X and s_{1n} is the determinant of the matrix that has row 1 and column n removed. The determinant of the matrix:

$$\begin{bmatrix} 1 & 2 & 3 \\ 4 & 5 & 6 \\ 7 & 8 & 0 \end{bmatrix}$$

is equal to:

$$1 \begin{vmatrix} 5 & 6 \\ 8 & 0 \end{vmatrix} - 2 \begin{vmatrix} 4 & 6 \\ 7 & 0 \end{vmatrix} + 3 \begin{vmatrix} 4 & 5 \\ 7 & 8 \end{vmatrix}$$

The next step is to evaluate the minors:

$$\begin{vmatrix} 5 & 6 \\ 8 & 0 \end{vmatrix} \quad \begin{vmatrix} 4 & 6 \\ 7 & 0 \end{vmatrix} \quad \begin{vmatrix} 4 & 5 \\ 7 & 8 \end{vmatrix}$$

according to the procedure:

$$1(5 \cdot 0 - 8 \cdot 6) - 2(4 \cdot 0 - 7 \cdot 6) + 3(4 \cdot 8 - 7 \cdot 5)$$

The resulting value of the determinant, in this case, is 27. If each element of a row or column is zero, then the determinant is zero. Also, if two rows or columns are identical, then the determinant is zero.

We have described the method for calculating the determinant of a matrix. Since we will be using determinants in later chapters, let us now consider a Pascal program that calculates determinants.

PASCAL PROGRAM: DETERMINANTS

A program that can be used to find the determinant of a 3-by-3 matrix is given in Figure 3.3. Procedure GET _ DATA asks the user to enter the matrix elements. Procedure DETER then calculates the determinant. Type up the program and run it. The elements of the matrix are entered row by row. That is, the order is:

A[1,1], A[1,2], A[1,3], A[2,1], A[2,2],...

After the nine elements have been entered, the program displays the value of the determinant. Then the question "More?" appears. Respond with a Y if you want to calculate the determinant of another matrix.

```
PROGRAM DETERM(INPUT, OUTPUT);
(* Feb 8, 81 *)


(* Pascal program to calculate *)
(* the determinant of a 3—by—3 matrix *)


TYPE
   ARY2 = ARRAY[1..3, 1..3] OF REAL;


VAR
   A     : ARY2;
   N     : INTEGER;
   YESNO : CHAR;
   D     : REAL;
```

Figure 3.3: The Determinant of a 3-by-3 Matrix

```
PROCEDURE GET _ DATA(VAR A : ARY2;
                     VAR N : INTEGER);
(* get values for n and arrays x, y *)

VAR
   I, J : INTEGER;

BEGIN
   N := 3;
   WRITELN;
   FOR I := 1 TO N DO
      BEGIN
         FOR J := 1 TO N DO
            BEGIN
               WRITE( J:3, ': ');
               READLN( A[I,J])
            END (* j loop *)
      END (* i loop *);
   WRITELN;
   FOR I:= 1 TO N DO
      BEGIN
         FOR J:= 1 TO N DO
            WRITE( A[I,J] :7:4, ' ');
         WRITELN
      END;
   WRITELN
END (* procedure get _ data *);

FUNCTION DETER (A : ARY2) : REAL;

(* calculate the determinant of a 3—by—3 matrix *)

VAR
   SUM : REAL;
```

Figure 3.3: The Determinant of a 3-by-3 Matrix (cont.)

```
BEGIN
  SUM := A[1,1] * (A[2,2]*A[3,3] — A[3,2]*A[2,3])
        — A[1,2] * (A[2,1]*A[3,3] — A[3,1]*A[2,3])
        + A[1,3] * (A[2,1]*A[3,2] — A[3,1]*A[2,2]);
  DETER := SUM
END;

BEGIN  (* main program *)
  REPEAT
    GET_DATA (A, N);
    D := DETER(A);
    WRITELN( 'The determinant is ', D);
    WRITELN;
    WRITE(' More? ');
    READLN( YESNO)
  UNTIL (YESNO <> 'Y' ) AND (YESNO <> 'y')
END.
```

Figure 3.3: The Determinant of a 3-by-3 Matrix (cont.)

INVERSE MATRICES AND MATRIX DIVISION

The *inverse* of a nonsingular matrix X is written as

$$X^{-1}$$

The inverse of a singular matrix is undefined. The product of a matrix and its inverse is the identity matrix

$$XX^{-1} = I$$

Matrix inversion is similar to other inverse operations since the inverse of an inverse produces the original matrix

$$(X^{-1})^{-1} = X$$

In computer implementations, however, the two will not always agree because of roundoff error.

Matrix division is performed by a combination of inversion and multiplication. Thus, if we need to divide matrix X by matrix Y, we first perform a matrix inversion on Y. Then matrix X is multiplied by the inverse of matrix Y. The operation is written as:

$$XY^{-1}$$

Matrix inversion can be used to find the solution to a set of simultaneous linear equations. Thus, if we have a coefficient matrix A and a constant vector $\mathbf{y}$, we want a solution vector $\mathbf{b}$ such that

$$A\mathbf{b} = \mathbf{y}$$

Then the solution can be obtained from a product of the inverse of matrix A and the constant vector $\mathbf{y}$ (in that order).

$$A^{-1}\mathbf{y} = \mathbf{b}$$

Since the simultaneous solution of linear equations is the subject of Chapter 4, we shall not develop a matrix inversion routine at this point.

SUMMARY

In this chapter we saw how vectors and matrices are represented in Pascal. We also studied the methods of performing matrix and vector arithmetic operations in Pascal. We developed two significant programs —one for performing matrix multiplication, and the other for calculating determinants. We will be using these programs in the chapters that follow.

CHAPTER **4**

Simultaneous Solution of Linear Equations

INTRODUCTION

In this chapter we will consider the simultaneous solution to a set of linear equations. We will use Cramer's rule, the Gauss elimination method, the Gauss-Jordan elimination method, and the Gauss-Seidel method. In addition, we will study the problem of *ill conditioning* by generating a set of Hilbert matrices. We will develop special Pascal application programs. These include the solution to a set of equations with multiple constant vectors, and equations with complex coefficients. We will also present a program for producing the best fit to an overdetermined system.

We will begin by describing linear equations and providing a simple example of simultaneous equations.

LINEAR EQUATIONS AND SIMULTANEOUS SOLUTIONS

A *linear equation* consists of a sum of terms such as:

$$Ax + By + Cz = D$$

In this equation, x, y and z are variables and A, B, C, and D are constants. No more than one variable can occur in each term, and it must be present to the first power. Thus, expressions such as:

$$2x^2 + 3y^2 = 4$$
$$\sin(x) + \log(y) = 9$$

and

$$xy = 2$$

are *nonlinear equations* if the variables are x and y.

An equation such as:

$$\frac{x}{A} + y\log(B) = p$$

is linear in the parameters x and y. On the other hand, if the symbols A and B are considered as the variables, and the symbols x, y and p are known values, then the equation is nonlinear.

Linear equations occur frequently in all branches of science and engineering and so effective methods are needed for solving them. For the particular case where there are several unknowns and an equal number of independent equations, then a unique solution is possible. In this chapter, we will consider several different methods for the solution of a set of linear equations. The solution of nonlinear equations is discussed in other chapters.

The two linear equations:

$$x - 2y = 1$$
$$2x + y = 7$$

represent two straight lines in the x-y plane. The simultaneous solution of these two equations represents the intersection of the two lines. Therefore, a graphical solution can be obtained by plotting the lines and finding the point of intersection.

The first equation has a y-intercept of -0.5 and a slope of 0.5. This can be seen by rearranging the equation as:

$$y = -0.5 + 0.5x$$

The second equation has an intercept of 7 and a slope of -2:

$$y = 7 - 2x$$

The intersection of the two lines occurs at the location:

$$x = 3, \quad y = 1$$

which represents the solution to the problem. This solution can be verified by substitution into the original equations.

Another method for finding the simultaneous solution to these two equations is to multiply the second equation by 2 and add it to the first equation. The resulting equation contains only the variable x, and so it can be solved directly.

$$
\begin{aligned}
x - 2y &= 1 \\
4x + 2y &= 14 \\
\hline
5x &= 15 \quad \text{or} \quad x = 3
\end{aligned}
$$

This value of x can then be substituted into either one of the original equations to find the corresponding value of y. Both of the above methods are suitable for solving two simultaneous equations. But, in general, these methods are tedious for larger numbers of equations. In the next section we will study a more sophisticated method of solving simultaneous equations, using matrices and vectors.

SOLUTION BY CRAMER'S RULE

A technique known as Cramer's rule is useful for solving two or three simultaneous equations. This is a particularly powerful method when the solution is performed by hand, or when a pocket calculator is used. In this approach, the equations are written with the unknowns on one side of the equal sign, and the constant terms on the other. Terms containing the same unknowns are vertically aligned. The two equations in the previous section were initially written in this form.

The coefficients of the unknowns are placed into a matrix known as the *coefficient matrix*. The constant terms are put into a separate vector. For our two equations, this produces the matrix A and the vector $\mathbf{z}$.

$$A = \begin{bmatrix} 1 & -2 \\ 2 & 1 \end{bmatrix} \qquad \mathbf{z} = \begin{bmatrix} 1 \\ 7 \end{bmatrix}$$

The solution is found from the relationship:

$$x = \frac{D_1}{D} \qquad y = \frac{D_2}{D}$$

where D is the determinant of the coefficient matrix:

$$D = \begin{vmatrix} 1 & -2 \\ 2 & 1 \end{vmatrix} = (1)(1) - (2)(-2) = 5$$

The determinant D_1 is found by substituting the constant vector into column 1 of the matrix. This column corresponds to the unknown x:

$$D_1 = \begin{vmatrix} 1 & -2 \\ 7 & 1 \end{vmatrix} = (1)(1) - (7)(-2) = 15$$

Similarly, D_2 is obtained by substituting the constant vector into column 2:

$$D_2 = \begin{vmatrix} 1 & 1 \\ 2 & 7 \end{vmatrix} = (1)(7) - (2)(1) = 5$$

The solution is then:

$$x = \frac{15}{5} = 3 \quad \text{and} \quad y = \frac{5}{5} = 1$$

Of course some sets of equations have no unique solutions. We will now examine how Cramer's Rule handles such equations.

Linear Dependence

We can see that if the determinant of the coefficient matrix is zero, then no unique solution is possible. This will occur whenever there is a *linear dependence* among the equations, that is, one of the equations can be obtained from a combination of one or more of the others. The matrix is said to be singular in this case. As an example of linear dependence, consider the two equations:

$$x + y = 5$$
$$2x + 2y = 2$$

The coefficient matrix for these two equations is:

$$\begin{bmatrix} 1 & 1 \\ 2 & 2 \end{bmatrix}$$

and the corresponding determinant is zero. These two equations represent parallel lines, which, of course, do not intersect. Consequently, there can be no solution.

As a second example, consider the two equations:

$$x + y = 5$$
$$2x + 2y = 10$$

This example produces the same singular coefficient matrix as the previous one. In this case, however, the two lines lie on top of each other, and so there are an infinite number of solutions.

Now that we have a powerful method of solving two or three simultaneous equations, let us consider a practical application of the method. We will solve the problem of the following section using a Pascal program we developed in Chapter 3.

Example: A Direct-Current Electrical Circuit

An interesting practical problem where simultaneous equations must be solved is provided by an electrical circuit. Consider, for example, the network of resistors and direct-current voltage sources shown in Figure 4.1. This circuit contains four nodes and six branches. There is a 20-volt source on the left side of the network, and a 5-volt source on the right side. In addition, there are six resistors of known value.

The problem is to find the resulting branch currents and the voltages across each resistor. The electrical currents in the six separate branches can be determined by solving six simultaneous equations. The program can be simplified, however, by considering three loop currents and the corresponding three simultaneous equations. Then, the branch currents can be found from the loop currents.

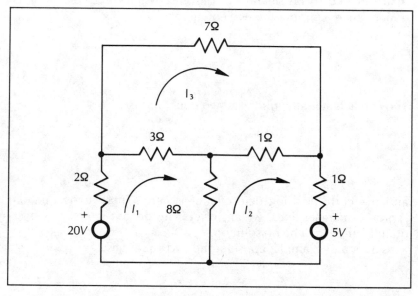

Figure 4.1: A Network of Resistors and DC Sources

The loop current in the lower-left loop is designated as I_1, the loop current in the lower-right loop is I_2 and the loop current in the upper loop is I_3. The three loop equations can be derived from the Kirchhoff voltage law by going around each loop in a counter-clockwise direction:

$$13I_1 - 8I_2 - 3I_3 - 20 = 0 \text{ (lower-left loop)}$$
$$-8I_1 + 10I_2 - I_3 + 5 = 0 \text{ (lower-right loop)}$$
$$-3I_1 - I_2 + 11I_3 = 0 \text{ (upper loop)}$$

The corresponding coefficient matrix and constant vector are:

$$\begin{bmatrix} 13 & -8 & -3 \\ -8 & 10 & -1 \\ -3 & -1 & 11 \end{bmatrix} \quad \begin{bmatrix} 20 \\ -5 \\ 0 \end{bmatrix}$$

Proceeding by Cramer's rule, we find:

$$D = \begin{vmatrix} 13 & -8 & -3 \\ -8 & 10 & -1 \\ -3 & -1 & 11 \end{vmatrix} \qquad D_1 = \begin{vmatrix} 20 & -8 & -3 \\ -5 & 10 & -1 \\ 0 & -1 & 11 \end{vmatrix}$$

$$D_2 = \begin{vmatrix} 13 & 20 & -3 \\ -8 & -5 & -1 \\ -3 & 0 & 11 \end{vmatrix} \qquad D_3 = \begin{vmatrix} 13 & -8 & 20 \\ -8 & 10 & -5 \\ -3 & -1 & 0 \end{vmatrix}$$

These determinants can be found by using the program DETERM given in Figure 3.3 of the previous chapter. The resulting loop currents are:

$$I_1 = \frac{D_1}{D} = \frac{1725}{575} = 3 \text{ amps}$$

$$I_2 = \frac{D_2}{D} = \frac{1150}{575} = 2 \text{ amps}$$

$$I_3 = \frac{D_3}{D} = \frac{575}{575} = 1 \text{ amp}$$

While program DETERM gives us the correct answer, using it is unnecessarily complicated. Nine separate numbers must be entered into the computer program for each of the four determinants. This requires the entry of thirty-six separate values.

We will now study an efficient program designed specifically to make use of Cramer's Rule for solving simultaneous equations.

PASCAL PROGRAM: A MORE ELEGANT USE OF CRAMER'S RULE

The program shown in Figure 4.2 simplifies the process considerably. The mathematical operations are exactly the same as in our first solution. But with this version, it is only necessary to enter twelve numbers: nine for the coefficients and three for the constant vector.

Running the Program

Type up the program shown in Figure 4.2 and execute it. The coefficients and constant term for the first equation are entered on the first line. The corresponding values for the second equation are entered on the second line and the data for the third equation are entered on the third line. Press the space bar after entering each coefficient, then press RETURN after entering the constant term. The input will look like this:

```
Simultaneous solution by Cramers rule

Equation   1
  1: 13    2: -8    3: -3 , C: 20
Equation   2
  1: -8    2: 10    3: -1 , C: -5
Equation   3
  1: -3    2: -1    3: 11 , C: 0
```

The data will be reprinted and the solution will follow:

```
13.0000   -8.0000   -3.0000    : 20.0000
-8.0000   10.0000   -1.0000    : -5.0000
-3.0000   -1.0000   11.0000    :  0.0000

 3.00000  2.00000  1.00000
```

Then, the question

```
More?
```

will appear. Answer with a Y if you want to solve another set. Otherwise enter an N.

The data can alternately be entered with one number on each line. In this case, the RETURN key is pressed after each number is given. The input will then look like this:

```
Equation   1
  1: 13
  2: -8
  3: -3
, C: 20
Equation   2
  1: -8
  2: 10
  3: -1
, C: -5
Equation   3
  1: -3
  2: -1
  3: 11
, C: 0
```

```
PROGRAM SIMQ1(INPUT,OUTPUT);
(* Feb 12, 81 *)
(* Pascal program to solve three *)
(* simultaneous equations by Cramer's rule *)

CONST
    RMAX = 3;
    CMAX = 3;

TYPE
    ARYS  = ARRAY[1..CMAX] OF REAL;
    ARY2S = ARRAY[1..RMAX, 1..CMAX] OF REAL;

VAR
    Y, COEF : ARYS;
    A        : ARY2S;
    N        : INTEGER;
    YESNO  : CHAR;
    ERROR  : BOOLEAN;

PROCEDURE GET_DATA(VAR A : ARY2S;
                    VAR Y : ARYS;
                    VAR N : INTEGER);

(* get values for n and arrays a, y *)

VAR
    I, J : INTEGER;

BEGIN (* procedure get_data *)
    WRITELN;
    N := RMAX;
```

Figure 4.2: Solution of Three Linear Equations by Cramer's Rule

```
    FOR I := 1 TO N DO
       BEGIN
          WRITELN( ' Equation ', I:3);
          FOR J := 1 TO N DO
             BEGIN
                WRITE( J:3, ': ');
                READ( A[I,J])
             END;
          WRITE( ', C: ');
          READLN (Y[I])
       END;
    WRITELN;
    FOR I:= 1 TO N DO
       BEGIN
          FOR J:= 1 TO N DO
             WRITE( A[I,J] :7:4, ' ');
          WRITELN( ' : ', Y[I] :7:4)
       END;
    WRITELN
END (* procedure get_data *);


PROCEDURE WRITE_DATA;
(* print out the answers *)

VAR
   I : INTEGER;

BEGIN (* write_data *)
   FOR I := 1 TO N DO
      WRITE( COEF[I] :9:5);
   WRITELN
END (* write_data *);
```

Figure 4.2: Solution of Three Linear Equations by Cramer's Rule (cont.)

```
PROCEDURE SOLVE( A : ARY2S;
                 Y : ARYS;
         VAR   COEF : ARYS;
                 N : INTEGER;
         VAR   ERROR : BOOLEAN);

VAR
    B    : ARY2S;
    I, J  : INTEGER;
    DET  : REAL;

FUNCTION DETER (A : ARY2S) : REAL;

(* calculate the determinant of a 3 — by — 3 matrix *)

VAR
    SUM : REAL;

BEGIN (* function deter *)
    SUM := A[1,1] * (A[2,2]*A[3,3] — A[3,2]*A[2,3])
         — A[1,2] * (A[2,1]*A[3,3] — A[3,1]*A[2,3])
         + A[1,3] * (A[2,1]*A[3,2] — A[3,1]*A[2,2]);
    DETER := SUM
END (* function deter *);

PROCEDURE SETUP (VAR   B : ARY2S;
                 VAR   COEF : ARYS;
                         J : INTEGER);

VAR
    I : INTEGER;
```

Figure 4.2: Solution of Three Linear Equations by Cramer's Rule (cont.)

```
BEGIN (* setup *)
   FOR I := 1 TO N DO
      BEGIN
         B[I,J] := Y[I];
         IF J > 1 THEN B[I,J−1] := A[I,J−1]
      END;
   COEF[J] := DETER(B)/DET
END (* setup *);


BEGIN (* procedure solve *)
   ERROR := FALSE;
   FOR I := 1 TO N DO
      FOR J := 1 TO N DO
         B[I,J] := A[I,J];
   DET := DETER(B);
   IF DET = 0.0 THEN
      BEGIN
         ERROR := TRUE;
         WRITELN(' ERROR: matrix singular')
      END
   ELSE
      BEGIN
         SETUP( B, COEF, 1);
         SETUP( B, COEF, 2);
         SETUP( B, COEF, 3)
      END (* else *)
END (* procedure solve *);
```

Figure 4.2: Solution of Three Linear Equations by Cramer's Rule (cont.)

```
BEGIN (* main program *)
    WRITELN;
    WRITELN(' Simultaneous solution by Cramers rule');
    REPEAT
        GET_DATA (A, Y, N);
        SOLVE (A, Y, COEF, N, ERROR);
        IF NOT ERROR THEN WRITE_DATA;
        WRITELN;
        WRITE(' More? ');
        READLN( YESNO)
    UNTIL (YESNO <> 'Y' ) AND (YESNO <> 'y')
END.
```

Figure 4.2: Solution of Three Linear Equations by Cramer's Rule (cont.)

SOLUTION BY GAUSS ELIMINATION

Cramer's rule is an effective method for solving two or three simultaneous equations, particularly when the solution is performed by hand. However, the computation time increases with the fourth power of the matrix size. Consequently, it will take about 16 times longer to solve six simultaneous equations than it does to solve three equations. The Gauss elimination method is more efficient. Computation time increases with the third power of the number of equations, and so it will take about 8 times longer to solve six equations than it does to solve three.

The Gauss elimination method can be readily programmed on a digital computer. However, the technique is fairly complicated, and therefore not suitable for hand solution. Furthermore, the operations involve frequent multiplication, division and subtraction. The consequent loss of precision places a practical upper limit on the number of equations that can be solved simultaneously. Let us go through the method step by step.

The Steps of the Gauss Method

With the Gauss elimination method, the original equations are manipulated so that the coefficient matrix contains a value of unity at each point on the major diagonal and a value of zero at each position below and to the left of the major diagonal.

Two basic types of matrix operations are used in the Gauss elimination method: scalar multiplication and addition. Any equation can be multiplied by a constant without changing the result. This is equivalent to multiplying one row of the coefficient matrix and the corresponding element in the constant vector by the same value. Also, any equation can be replaced by the sum of two equations.

The following steps will demonstrate the use of Gauss elimination using the equations we derived from the electric circuit shown in Figure 4.1. Initially, the coefficient matrix and the constant vector are:

$$\begin{bmatrix} 13 & -8 & -3 \\ -8 & 10 & -1 \\ -3 & -1 & 11 \end{bmatrix} \quad \begin{bmatrix} 20 \\ -5 \\ 0 \end{bmatrix}$$

The first variable is eliminated from all but the first equation. The equations are manipulated to produce the value of unity at the top of the first column. The remaining positions of the column become zeros. The operation is performed in the following way. The entire first row is divided by the first element in the row (the pivot element). This generates the value of unity in the first diagonal position. By this means, the first equation becomes:

$$\begin{bmatrix} 1 & -0.61 & -0.23 \end{bmatrix} \quad \begin{bmatrix} 1.5 \end{bmatrix}$$

The first unknown is eliminated from the second row by combining the first two rows. The new first row is multiplied by the first element in the second row, then subtracted from the second row. The new second row is:

$$\begin{bmatrix} 0 & 5.1 & -2.8 \end{bmatrix} \quad \begin{bmatrix} 7.3 \end{bmatrix}$$

In a similar fashion, the first variable is eliminated from the third equation. The three equations are now approximately:

$$\begin{bmatrix} 1 & -0.61 & -0.23 \\ 0 & 5.1 & -2.8 \\ 0 & -2.8 & 10.3 \end{bmatrix} \quad \begin{bmatrix} 1.5 \\ 7.3 \\ 4.6 \end{bmatrix}$$

The next step is to produce the value of unity in the second position of the second row (the new pivot). The second line is divided by the second element. The second variable is eliminated from the third equation by generating a zero in the second position, just under the pivot element. The three equations now look like:

$$\begin{bmatrix} 1 & -0.61 & -0.23 \\ 0 & 1 & -0.56 \\ 0 & 0 & 8.7 \end{bmatrix} \quad \begin{bmatrix} 1.5 \\ 1.4 \\ 8.7 \end{bmatrix}$$

The final step of this phase is to obtain a value of unity in the third pivot position. This is accomplished by dividing the third equation by the pivot value to give:

$$\begin{bmatrix} 1 & -0.61 & -0.23 \\ 0 & 1 & -0.56 \\ 0 & 0 & 1 \end{bmatrix} \quad \begin{bmatrix} 1.5 \\ 1.4 \\ 1 \end{bmatrix}$$

The result corresponds to the three equations:

$$\begin{aligned} x - 0.61y - 0.23z &= 1.5 \\ y - 0.56z &= 1.4 \\ z &= 1 \end{aligned}$$

The third equation can be solved directly since it has only one unknown. The result is:

$$z = 1$$

The second equation becomes:

$$y = 1.4 + 0.56z$$

By substituting the value of z into this equation, the value of y is found to be 2. Substituting the values of y and z into the first equation produces the value of 3 for x. This phase of the calculations is known as *back substitution*.

Improving the Accuracy of the Gauss Method

The accuracy of the Gauss elimination method can be improved by interchanging two rows so that the element with the largest absolute magnitude becomes the pivot element. Suppose, for example, that the previous three equations had been originally written in a different order:

$$\begin{bmatrix} -3 & -1 & 11 \\ 13 & -8 & -3 \\ -8 & 10 & -1 \end{bmatrix} \quad \begin{bmatrix} 0 \\ 20 \\ -5 \end{bmatrix}$$

Then the top equation could be divided by -3 to put the value of unity in the pivot position. However, the result will be more accurate if the first and second rows are interchanged first. This will put the larger value of 13 in the pivot position. After the first variable is eliminated from the second and third equations, the result is:

$$\begin{bmatrix} 1 & -0.61 & -0.23 \\ 0 & -2.8 & 10.3 \\ 0 & 5.1 & -2.8 \end{bmatrix} \quad \begin{bmatrix} 1.5 \\ 4.6 \\ 7.3 \end{bmatrix}$$

Again, it would be best to interchange the second and third equations to put the larger element of 5.1 in the second pivot position.

There is another reason why rows may have to be interchanged. If a zero element appears on the major diagonal, then it will not be possible to divide the row by this pivot element. But interchanging this row with one that is below it will remove the zero from the pivot position.

Now that we have gone through the rather tedious process of using Gaussian elimination to solve equations by hand, we can appreciate the elegance of the Pascal program in the following section.

PASCAL PROGRAM: THE GAUSS ELIMINATION METHOD

The program shown in Figure 4.3 can be used to simultaneously solve a set of linear equations by Gaussian elimination. The program is written to handle up to eight equations. The size can be increased by changing the two variables called MAXR and MAXC near the beginning. Type up the program and execute it.

```
PROGRAM GAUS(INPUT, OUTPUT);
(* Feb 12, 81 *)

(* Pascal program to perform *)
(* simultaneous solution by Gaussian elimination *)
(* procedure GAUSS is included *)


CONST
   MAXR = 8;
   MAXC = 8;


TYPE
   ARY    = ARRAY[1..MAXR] OF REAL;
   ARYS   = ARRAY[1..MAXC] OF REAL;
   ARY2S = ARRAY[1..MAXR, 1..MAXC] OF REAL;


VAR
   Y      : ARYS;
   COEF   : ARYS;
   A      : ARY2S;
   N, M   : INTEGER;
   ERROR : BOOLEAN;


PROCEDURE GET _ DATA(VAR A : ARY2S;
                     VAR Y : ARYS;
                     VAR N, M : INTEGER);


(* get values for n and arrays a, y *)

VAR
   I, J : INTEGER;
```

Figure 4.3: Simultaneous Solution by Gauss Elimination

```
BEGIN
  WRITELN;
  REPEAT
    WRITE( ' How many equations? ');
    READLN( N);
    M := N
  UNTIL N < MAXR;
  IF N > 1 THEN
    BEGIN
      FOR I := 1 TO N DO
        BEGIN
          WRITELN( ' Equation', I:3);
          FOR J := 1 TO N DO
            BEGIN
              WRITE( J:3, ': ');
              READ( A[I,J])
            END;
          WRITE( ', C: ');
          READ (Y[I]);
          READLN (* clear line *)
        END;
      WRITELN;
      FOR I := 1 TO N DO
        BEGIN
          FOR J := 1 TO M DO
            WRITE( A[I,J] :7:4, ' ');
          WRITELN( ' : ', Y[I] :7:4)
        END;
      WRITELN
    END (* if n>1 *)
END (* procedure get_data *);
```

Figure 4.3: Simultaneous Solution by Gauss Elimination (cont.)

```
PROCEDURE WRITE_DATA;
(* print out the answers *)

VAR
   I : INTEGER;

BEGIN
   FOR I := 1 TO M DO
      WRITE( COEF[I] :9:5);
   WRITELN
END (* write_data *);

PROCEDURE GAUSS
               (A       : ARY2S;
                Y       : ARYS;
            VAR  COEF : ARYS;
                NCOL : INTEGER;
            VAR  ERROR : BOOLEAN);

(* matrix solution by Gaussian Elimination *)
(* Feb 8, 81 *)
(* Adapted from Gilder *)

VAR
   B     : ARY2S (* work array, nrow,ncol *);
   W     : ARYS  (* work array, ncol long *);
   I, J, I1, K, L,
   N     : INTEGER;
   HOLD, SUM, T, AB, BIG : REAL;

BEGIN
   ERROR := FALSE;
   N := NCOL;
```

Figure 4.3: Simultaneous Solution by Gauss Elimination (cont.)

```
FOR I := 1 TO N DO
   BEGIN (* copy to work arrays *)
      FOR J := 1 TO N DO
         B[I,J] := A[I,J];
      W[I] := Y[I]
   END;
FOR I := 1 TO N − 1 DO
   BEGIN
      BIG := ABS(B[I,I]);
      L := I;
      I1 := I + 1;
      FOR J := I1 TO N DO
         BEGIN (* search for largest element *)
            AB := ABS(B[J,I]);
            IF AB > BIG THEN
               BEGIN
                  BIG := AB;
                  L := J
               END
         END;
      IF BIG = 0.0 THEN ERROR := TRUE
      ELSE
         BEGIN
            IF L <> I THEN
               BEGIN
                  (* interchange rows to put *)
                  (* largest element on diagonal *)
                  FOR J := 1 TO N DO
                     BEGIN
                        HOLD := B[L,J];
                        B[L,J] := B[I,J];
                        B[I,J] := HOLD
                     END;
```

Figure 4.3: Simultaneous Solution by Gauss Elimination (cont.)

```
                    HOLD := W[L];
                    W[L] := W[I];
                    W[I] := HOLD
                  END (* if L <> i *);
                FOR J := I1 TO N DO
                BEGIN
                  T := B[J,I]/B[I,I];
                  FOR K := I1 TO N DO
                    B[J,K] := B[J,K] − T * B[I,K];
                  W[J] := W[J] − T * W[I]
                END (* j loop *)
              END (* if big *)
          END; (* i loop *)
      IF B[N,N] = 0.0 THEN ERROR := TRUE
      ELSE
        BEGIN
          COEF[N] := W[N]/B[N,N];
          I := N − 1;
          (* back substitution *)
          REPEAT
            SUM := 0.0;
            FOR J := I+1 TO N DO
              SUM := SUM + B[I,J] * COEF[J];
            COEF[I] := (W[I] − SUM)/B[I,I];
            I := I − 1
          UNTIL I = 0
        END (* IF b[n,n] = 0 *);
      IF ERROR THEN WRITELN('ERROR: Matrix singular ')
  END (* gauss *);


BEGIN (* main program *)
  WRITELN;
  WRITELN(' Simultaneous solution by Gauss elimination');
```

Figure 4.3: Simultaneous Solution by Gauss Elimination (cont.)

```
REPEAT
    GET_DATA (A, Y, N, M);
    IF N > 1 THEN
        BEGIN
            GAUSS (A, Y, COEF, N, ERROR);
            IF NOT ERROR THEN WRITE_DATA
        END
    UNTIL N < 2
END.
```

Figure 4.3: Simultaneous Solution by Gauss Elimination (cont.)

Running the Program

This program is similar to the previous one. But it begins by asking for the number of equations. Answer this question with a number from 2 to 8, then press the carriage return. The coefficients for each equation and the corresponding constant vectors are entered in turn. Put one equation on each line. Press the space bar after each number is entered. The constant term is then added at the end of the line. Press the carriage return to finish the line. Enter the values for the set of equations we considered earlier in this chapter and verify that the solution vector is [3 2 1]. The results should look like Figure 4.4.

```
Simultaneous solution by Gauss elimination

How many equations? 3
Equation  1
  1: 13    2: -8    3: -3  , C: 20
Equation  2
  1: -8    2: 10    3: -1  , C: -5
Equation  3
  1: -3    2: -1    3: 11  , C: 0

 13.0000   -8.0000   -3.0000   :  20.0000
 -8.0000   10.0000   -1.0000   :  -5.0000
 -3.0000   -1.0000   11.0000   :   0.0000

  3.00000   2.00000   1.00000
```

Figure 4.4: Output: Solution of the Electrical Circuit Problem Using Gauss Elimination

At the completion of the task, the program begins again. This time find the simultaneous solution to the following three equations and write down the answer for later reference.

$$\begin{bmatrix} 1 & 1 & 1 \\ 2 & 1 & -1 \\ 3 & 1 & -3 \end{bmatrix} \qquad \begin{bmatrix} 6 \\ 1 \\ -4 \end{bmatrix}$$

We will come back to this set at the end of the next section.

For the third task, enter two lines that are exactly the same, for example:

$$\begin{bmatrix} 1 & 1 & 1 \\ 1 & 1 & 1 \\ 2 & 1 & -1 \end{bmatrix} \qquad \begin{bmatrix} 6 \\ 6 \\ 1 \end{bmatrix}$$

The program should print an error message indicating that the matrix is singular. The program can be aborted by entering a value less than two for the number of equations.

SOLUTION BY GAUSS-JORDAN ELIMINATION

A variation of the Gauss elimination method is known as the Gauss-Jordan method. The approach shares most of the advantages and disadvantages of the Gauss elimination technique. Execution time is a third-order function of the matrix size and there are many multiplication, division, and subtraction operations that contribute to loss of accuracy. Furthermore, the Gauss-Jordan algorithm is more complicated than the one for Gauss elimination. Nevertheless, the Gauss-Jordan technique will generally be the most useful of all. In fact, we will use it in later chapters of this book. Its usefulness lies in the fact that the inverse of the coefficient matrix is readily obtained along with the solution vector. Let us outline the steps of this method.

Details of the Gauss-Jordan Method

In the Gauss-Jordan method, the elements of the major diagonal are converted to unity as with the Gauss method. But now the elements both above and below the major diagonal are converted to zeros. Thus, the coefficient matrix is converted to a unit matrix. The resulting constant vector then becomes the solution vector. For the Gauss-Jordan solution of the electrical circuit given in Figure 4.1, the final set of values becomes:

$$
\begin{bmatrix} 1 & 0 & 0 \\ 0 & 1 & 0 \\ 0 & 0 & 1 \end{bmatrix} \quad \begin{bmatrix} 3 \\ 2 \\ 1 \end{bmatrix}
$$

This corresponds to the three equations:

$$
\begin{aligned}
x & & & = 3 \\
& y & & = 2 \\
& & z & = 1
\end{aligned}
$$

for which the solution is:

$$
x = 3, \quad y = 2, \quad z = 1
$$

Suppose that a unit matrix is initially placed to the right of the original set of equations. Then, if all of the operations are performed on this matrix as they are performed on the other elements, this unit matrix will be converted into the inverse of the original coefficient matrix at the conclusion of the calculation. That is, the set:

$$
\begin{bmatrix} 13 & -8 & -3 \\ -8 & 10 & -1 \\ -3 & -1 & 11 \end{bmatrix} \quad \begin{bmatrix} 20 \\ -5 \\ 0 \end{bmatrix} \quad \begin{bmatrix} 1 & 0 & 0 \\ 0 & 1 & 0 \\ 0 & 0 & 1 \end{bmatrix}
$$

will be transformed into:

$$
\begin{bmatrix} 1 & 0 & 0 \\ 0 & 1 & 0 \\ 0 & 0 & 1 \end{bmatrix} \quad \begin{bmatrix} 3 \\ 2 \\ 1 \end{bmatrix} \quad \begin{bmatrix} 0.19 & 0.16 & 0.07 \\ 0.16 & 0.23 & 0.06 \\ 0.07 & 0.06 & 0.11 \end{bmatrix}
$$

In this case, the matrix on the right is the inverse of the original coefficient matrix.

However, it is not actually necessary to use two separate matrices for these operations. The inverse matrix can be physically generated in the same space occupied by the original coefficient matrix. At the conclusion of the operation, the inverse matrix will then be returned in place of the coefficient matrix. And, of course, the solution vector will appear in place of the original constant vector. Thus, if the coefficient matrix and constant vector:

$$\begin{bmatrix} 13 & -8 & -3 \\ -8 & 10 & -1 \\ -3 & -1 & 11 \end{bmatrix} \qquad \begin{bmatrix} 20 \\ -5 \\ 0 \end{bmatrix}$$

are passed to the Gauss-Jordan routine, the matrix inverse and solution vector can be returned in the same array space:

$$\begin{bmatrix} 0.19 & 0.16 & 0.07 \\ 0.16 & 0.23 & 0.06 \\ 0.07 & 0.06 & 0.11 \end{bmatrix} \qquad \begin{bmatrix} 3 \\ 2 \\ 1 \end{bmatrix}$$

We will now present a Pascal program that implements this rather complex algorithm.

PASCAL PROGRAM: GAUSS-JORDAN ELIMINATION

The program shown in Figure 4.5 will solve simultaneous linear equations by the method of Gauss-Jordan elimination. The main program calls procedure GAUSSJ, which is shown as a separate, external procedure. If your Pascal compiler does not allow external procedures, then you will have to include the procedure in the main program.

Procedure GAUSSJ is given in Figure 4.6. With this version, the inverse of the coefficient matrix is returned in place of the original matrix, B. This is why the original matrix A is duplicated in the main program prior to calling procedure GAUSSJ. On the other hand, the data vector, Y, is isolated from the solution vector, COEF.

```
PROGRAM SOLVGJ(INPUT, OUTPUT);
(* Feb 12, 81 *)

(* Pascal program to perform simultaneous solution *)
(* by Gauss—Jordan elimination *)

CONST
    MAXR = 8;
    MAXC = 8;

TYPE
    ARY   = ARRAY[1..MAXR] OF REAL;
    ARYS  = ARRAY[1..MAXC] OF REAL;
    ARY2S = ARRAY[1..MAXR, 1..MAXC] OF REAL;

VAR
    Y         : ARYS;
    COEF      : ARYS;
    A, B      : ARY2S;
    N, M, I, J: INTEGER;
    ERROR     : BOOLEAN;

PROCEDURE GET_DATA(VAR A : ARY2S;
                   VAR Y : ARYS;
                   VAR N, M : INTEGER);

(* get values for n and arrays a, y *)

VAR
    I, J : INTEGER;
```

Figure 4.5: Solution of Simultaneous Equations by Gauss-Jordan Elimination

```
BEGIN
  WRITELN;
  REPEAT
    WRITE( ' How many equations? ');
    READLN( N);
    M := N
  UNTIL N < MAXR;
  IF N > 1 THEN
    BEGIN
      FOR I := 1 TO N DO
        BEGIN
          WRITELN( ' Equation', I:3);
          FOR J := 1 TO N DO
            BEGIN
              WRITE( J:3, ': ');
              READ( A[I,J])
            END;
          WRITE( ', C: ');
          READLN (Y[I]) (* clear line *)
        END;
      WRITELN;
      FOR I := 1 TO N DO
        BEGIN
          FOR J := 1 TO M DO
            WRITE( A[I,J] :7:4, ' ');
          WRITELN( ' : ', Y[I] :7:4)
        END;
      WRITELN
    END (* if n > 1 *)
END (* procedure get_data *);


PROCEDURE WRITE_DATA;


(* print out the answers *)
```

Figure 4.5: Solution of Simultaneous Equations by Gauss-Jordan Elimination (cont.)

```
VAR
   I : INTEGER;

BEGIN
   FOR I := 1 TO M DO
      WRITE( COEF[I] :9:5);
   WRITELN
END (* write_data *);

(* PROCEDURE gaussj
   (VAR b        : ary2s;
    y            : arys;
    VAR coef     : arys;
    ncol         : integer;
    VAR error    : boolean);
extern; *)
(*$F GAUSSJ.PAS *)

BEGIN (* main program *)
   WRITELN;
   WRITELN
      (' Simultaneous solution by Gauss—Jordan elimination');
   REPEAT
      GET_DATA (A, Y, N, M);
      IF N > 1 THEN
         BEGIN
            FOR I := 1 TO N DO
               FOR J := 1 TO N DO
                  B[I,J] := A[I,J] (* setup work array *);
            GAUSSJ (B, Y, COEF, N, ERROR);
            IF NOT ERROR THEN WRITE_DATA
         END
   UNTIL N < 2
END.
```

Figure 4.5: Solution of Simultaneous Equations by Gauss-Jordan Elimination (cont.)

```
PROCEDURE GAUSSJ
    (VAR B      : ARY2S; (* square matrix of coefficients *)
     Y          : ARYS; (* constant vector *)
     VAR COEF   : ARYS; (* solution vector *)
     NCOL       : INTEGER; (* order of matrix *)
     VAR ERROR : BOOLEAN); (* true if matrix singular *)


(*   Gauss Jordan matrix inversion and solution *)
(*   Adapted from McCormick *)
(*   Feb 8, 81 *)
(*    B(N,N) coefficient matrix, becomes inverse *)
(*    Y(N) original constant vector *)
(*    W(N,M) constant vector(s) become solution vector *)
(*    DETERM is the determinant *)
(*    ERROR = 1 if singular *)
(*    INDEX(N,3) *)
(*    NV is number of constant vectors *)

LABEL 99;

VAR
    W      : ARRAY[1..MAXC, 1..MAXC] OF REAL;
    INDEX : ARRAY[1..MAXC, 1..3] OF INTEGER;
    I, J, K, L, NV, IROW, ICOL, N, L1      : INTEGER;
    DETERM, PIVOT, HOLD, SUM, T, AB, BIG : REAL;


PROCEDURE SWAP(VAR A, B: REAL);


VAR
    HOLD: REAL;
```

Figure 4.6: Gauss-Jordan Procedure

```
BEGIN (* swap *)
   HOLD := A;
   A := B;
   B := HOLD
END (* procedure swap *);

PROCEDURE GAUSJ2;

LABEL 98;

VAR
   I, J, K, L, L1: INTEGER;

PROCEDURE GAUSJ3;

VAR
   L: INTEGER;

BEGIN (* procedure gausj3 *)
   (* interchange rows to put pivot on diagonal *)
   IF IROW <> ICOL THEN
      BEGIN
         DETERM := − DETERM;
         FOR L := 1 TO N DO
            SWAP(B[IROW, L], B[ICOL, L]);
         IF NV > 0 THEN
            FOR L := 1 TO NV DO
               SWAP(W[IROW, L], W[ICOL, L])
      END (* if irow <> icol *)
END (* gausj3 *);
```

Figure 4.6: Gauss-Jordan Procedure (cont.)

```
BEGIN (* procedure gausj2 *)
  (* actual start of gaussj *)
  ERROR := FALSE;
  NV := 1 (* single constant vector *);
  N := NCOL;
  FOR I := 1 TO N DO
    BEGIN
      W[I, 1] := Y[I] (* copy constant vector *);
      INDEX[I, 3] := 0
    END;
  DETERM := 1.0;
  FOR I := 1 TO N DO
    BEGIN
      (* search for largest element *)
      BIG := 0.0;
      FOR J := 1 TO N DO
        BEGIN
          IF INDEX[J, 3] <> 1 THEN
            BEGIN
              FOR K := 1 TO N DO
                BEGIN
                  IF INDEX[K, 3] > 1 THEN
                    BEGIN
                      WRITELN(' ERROR: matrix singular');
                      ERROR := TRUE;
                      GOTO 98        (* abort *)
                    END;
                  IF INDEX[K, 3] < 1 THEN
                    IF ABS(B[J, K]) > BIG THEN
```

Figure 4.6: Gauss-Jordan Procedure (cont.)

```
                        BEGIN
                            IROW := J;
                            ICOL := K;
                            BIG := ABS(B[J, K])
                        END
                    END (* k loop *)
                END
        END (* j loop *);

    INDEX[ICOL, 3] := INDEX[ICOL, 3] + 1;
    INDEX[I, 1] := IROW;
    INDEX[I, 2] := ICOL;
    GAUSJ3 (* further subdivision of gaussj *);
    (* divide pivot row by pivot column *)
    PIVOT := B[ICOL, ICOL];
    DETERM := DETERM * PIVOT;
    B[ICOL, ICOL] := 1.0;

    FOR L := 1 TO N DO
        B[ICOL, L] := B[ICOL, L]/PIVOT;
    IF NV > 0 THEN
        FOR L := 1 TO NV DO
            W[ICOL, L] := W[ICOL, L]/PIVOT;
    (* reduce nonpivot rows *)

    FOR L1 := 1 TO N DO
        BEGIN
            IF L1 <> ICOL THEN
                BEGIN
                    T := B[L1, ICOL];
                    B[L1, ICOL] := 0.0;
                    FOR L := 1 TO N DO
                        B[L1, L] := B[L1, L] — B[ICOL, L] * T;
                    IF NV > 0 THEN
```

Figure 4.6: Gauss-Jordan Procedure (cont.)

```
                          FOR L := 1 TO NV DO
                              W[L1, L] := W[L1, L] — W[ICOL, L] * T;
                          END (* IF l1 <> icol *)
                  END
              END (* i loop *);
      98:
      END (* gausj2 *);


      BEGIN       (* Gauss—Jordan main program *)
          GAUSJ2 (* first half of gaussj *);
          IF ERROR THEN GOTO 99;
          (* interchange columns *)
          FOR I := 1 TO N DO
              BEGIN
                  L := N — I + 1;
                  IF INDEX[L, 1] <> INDEX[L, 2] THEN
                      BEGIN
                          IROW := INDEX[L, 1];
                          ICOL := INDEX[L, 2];
                          FOR K := 1 TO N DO
                              SWAP(B[K, IROW], B[K, ICOL])
                      END (* if index *)
              END (* i loop *);
          FOR K := 1 TO N DO
              IF INDEX[K, 3] <> 1 THEN
                  BEGIN
                      WRITELN(' ERROR: matrix singular');
                      ERROR := TRUE;
                      GOTO 99   (* abort *)
                  END;
          FOR I := 1 TO N DO
              COEF[I] := W[I, 1];
      99:
      END (* procedure gaussj *);
```

Figure 4.6: Gauss-Jordan Procedure (cont.)

Type up the program and execute it. Try out the electrical circuit equations and verify that the solution is [3 2 1]. Then try the set:

$$\begin{bmatrix} 1 & 1 & 1 \\ 2 & 1 & -1 \\ 3 & 1 & -3 \end{bmatrix} \qquad \begin{bmatrix} 6 \\ 1 \\ -4 \end{bmatrix}$$

Notice that the answer, in this case, is not the same as the one found by the Gauss method. Furthermore, the set:

$$x = 1, \quad y = 2, \quad z = 3$$

is also a solution. How can there be more than one solution to a set of linear equations? The answer is that one of the equations is a linear combination of the other two. The third equation, in this case, is equal to twice the second equation minus the first equation. Thus the coefficient matrix is singular. Actually, if the above three equations are tried on the Cramer's rule program given at the beginning of this chapter, the singularity will be readily found.

Unfortunately, this kind of linear dependence is difficult to find. The determinant of the coefficient matrix may be small, but it does not equal zero because of roundoff errors that accumulate during the elimination process. Fortunately, different algorithms for solving simultaneous equations will usually give different answers in this case. Consequently, if there is any question of linear dependence, the equations should be solved by at least two different methods.

We have seen how the Gauss-Jordan elimination program works for one coefficient matrix and its corresponding constant vector. Now we will explore ways of using and refining this program to deal conveniently with *multiple* constant vectors. To illustrate this situation we will return to the electrical circuit example that we set up earlier in this chapter. We will also continue discussion of inverse coefficient matrices and their use in solving simultaneous equations.

MULTIPLE CONSTANT VECTORS AND MATRIX INVERSION

In the previous chapter, we mentioned that the solution to a set of linear equations could be obtained by multiplying the inverse of the coefficient matrix by the constant vector. For example, if the coefficient matrix is A and the constant vector is **y**, then the solution vector **b** is:

$$\mathbf{b} = A^{-1} \, \mathbf{y}$$

But the methods developed in this chapter (Cramer's rule, Gauss elimination and Gauss-Jordan elimination) obtain the solution to a set of linear equations by direct methods. The inverse of the coefficient matrix is not utilized.

Sometimes we need to solve several sets of simultaneous equations that all have the same coefficient matrix but different constant vectors. In this case, we can invert the coefficient matrix. Then the separate solution vectors can be obtained from the product of the inverted matrix and each constant vector. Even in this case, however, it will generally be faster to perform a Gauss-Jordan elimination on the matrix and all of the constant vectors simultaneously. In fact, the Gauss-Jordan procedure given in the previous section is actually programmed for multiple constant vectors.

Consider, for example, the electrical circuit shown in Figure 4.1. Suppose that we would like to determine the loop currents for three different circuit configurations:

1. the original circuit
2. the circuit with the 5-volt source reversed
3. the circuit with both voltage sources reversed

These three different configurations correspond to the following three sets of equations:

$$\begin{bmatrix} 13 & -8 & -3 \\ -8 & 10 & -1 \\ -3 & -1 & 11 \end{bmatrix} \quad \begin{bmatrix} 20 \\ -5 \\ 0 \end{bmatrix}$$

$$\begin{bmatrix} 13 & -8 & -3 \\ -8 & 10 & -1 \\ -3 & -1 & 11 \end{bmatrix} \quad \begin{bmatrix} 20 \\ 5 \\ 0 \end{bmatrix}$$

$$\begin{bmatrix} 13 & -8 & -3 \\ -8 & 10 & -1 \\ -3 & -1 & 11 \end{bmatrix} \quad \begin{bmatrix} -20 \\ 5 \\ 0 \end{bmatrix}$$

All three sets of equations can be solved separately using one of the preceding techniques. In this case, however, it is more efficient to solve all three conditions at once by using the Gauss-Jordan procedure. The coefficient matrix contains the values common to the three different circuits. However, the constant vector now becomes a *constant matrix*. Each column of the constant matrix represents a different circuit configuration and will produce the corresponding solution. The actual matrices entered into the Gauss-Jordan routine are:

$$\begin{bmatrix} 13 & -8 & -3 \\ -8 & 10 & -1 \\ -3 & -1 & 11 \end{bmatrix} \quad \begin{bmatrix} 20 & 20 & -20 \\ -5 & 5 & 5 \\ 0 & 0 & 0 \end{bmatrix}$$

The Gauss-Jordan routine returns the following values to the calling program:

$$\begin{bmatrix} 0.19 & 0.16 & 0.07 \\ 0.16 & 0.23 & 0.06 \\ 0.07 & 0.06 & 0.11 \end{bmatrix} \quad \begin{bmatrix} 3 & 4.58 & -3 \\ 2 & 4.33 & -2 \\ 1 & 1.64 & -1 \end{bmatrix}$$

The left matrix, which originally contained the coefficients, now holds the inverse of the coefficient matrix. The right matrix, which initially contained the constant vectors, now contains the corresponding solution vectors. Thus, for the three separate circuits, the answers are:

		Solution	
Circuit	I_1	I_2	I_3
1	3	2	1
2	4.58	4.33	1.64
3	-3	-2	-1

In the previous version of the Gauss-Jordan program, we copied the original coefficient matrix A into a work array B before calling the Gauss-Jordan procedure. Also, at the beginning of the Gauss-Jordan routine, the incoming constant vector Y was copied into the first column of the constant matrix W. Then at the end of the Gauss-Jordan procedure, the solution from the first column of matrix W was copied into the solution vector COEF. Now we will look at a refined version of this program.

PASCAL PROGRAM:
GAUSS-JORDAN ELIMINATION, VERSION TWO

An alternate version of the Gauss-Jordan program is given in Figure 4.8. The formal parameters of the Gauss-Jordan procedure have been changed. As before, the inverse of the coefficient matrix is returned in the array that held the original coefficient matrix. But the solution matrix is now returned in the array that originally held the matrix of constant vectors. In addition, the determinant of the coefficient matrix is returned as a separate parameter.

Running the Program

Type up the program and execute it. As before, the user will be asked for the number of equations. Next will be a question about the number of constant vectors. If this question is answered with the value of unity, then the program will behave exactly as before. However, the solution vector is now printed vertically rather than horizontally.

With this version, the number of constant vectors may be greater than one. In this case, the constants for each equation are entered on the same line as the corresponding coefficients. For example, if the above three electrical circuits are collectively solved with this program, then the first line of data will be:

$$13 \ -8 \ -3 \ 20 \ 20 \ -20$$

Solve all three of the above electric circuits at the same time and verify that the correct answers are obtained.

This new version has another option. The number of constant vectors may be zero. In this case, only a matrix of coefficients is entered. The program will then print the inverse of the coefficient matrix. The determinant of the coefficient matrix will also be displayed.

Enter the coefficient matrix we previously had trouble with:

$$\begin{bmatrix} 1 & 1 & 1 \\ 2 & 1 & -1 \\ 3 & 1 & -3 \end{bmatrix}$$

This matrix is singular and so the determinant is zero. However, this Gauss-Jordan program may return a small, but nonzero, value because of roundoff error. A typical run might look like Figure 4.7. (Enter zero to abort the program.)

```
        Simultaneous solution by Gauss-Jordan
        Multiple constant vectors, or matrix inverse

        How many equations? 3
        How many constant vectors? 0
        Equation  1
         1: 1    2: 1    3: 1
        Equation  2
         1: 2    2: 1    3: -1
        Equation  3
         1: 3    2: 1    3: -3

                 Matrix
        1.0000    1.0000    1.0000
        2.0000    1.0000   -1.0000
        3.0000    1.0000   -3.0000

           Inverse
        -11184800.0     22369600.0    -11184800.0
         16777200.0    -33554400.0     16777200.0
         -5592400.0     11184800.0     -5592410.0

        Determinant  is   1.78814e-7

        How many equations? 0
```

Figure 4.7: Output from Second Version of the Gauss-Jordan Method

```pascal
PROGRAM SOLVGV(INPUT, OUTPUT);
 (* Feb 8, 81 *)
(* Pascal program to perform simultaneous solution *)
(* by Gauss — Jordan elimination *)
(* with multiple constant vectors *)

CONST
    MAXR = 7;
    MAXC = 7;

TYPE
    ARY2S = ARRAY[1..MAXR, 1..MAXC] OF REAL;

VAR
    A, Y      : ARY2S;
    N, NVEC  : INTEGER;
    ERROR     : BOOLEAN;
    DETERM   : REAL;

PROCEDURE GET _ DATA(VAR A : ARY2S;
                     VAR Y : ARY2S;
                     VAR N, NVEC : INTEGER);
(* get values for n, nvec, and arrays a, y *)

VAR
    I, J : INTEGER;

BEGIN
    WRITELN;
    REPEAT
      WRITE( ' How many equations? ');
      READLN( N)
    UNTIL N < MAXR;
```

Figure 4.8: Solution of Simultaneous Equations and Matrix Inversion by the Gauss-Jordan Method (Multiple constant vectors may be entered)

```
IF N > 1 THEN
    BEGIN
        WRITE( ' How many constant vectors? ');
        READLN( NVEC);
        FOR I : = 1 TO N DO
            BEGIN
                FOR J : = 1 TO N DO
                    BEGIN
                        WRITE(', ', J:3, ': ');
                        READ( A[I,J])
                    END;
                IF NVEC > 0 THEN
                    BEGIN
                        FOR J : = 1 TO NVEC DO
                            BEGIN
                                WRITE( ', C: ');
                                READ(Y[I,J])
                            END;
                        READLN
                    END
            END (* I loop *);
        WRITELN;
        WRITE( '         Matrix');
        IF NVEC > 0 THEN WRITE('         Constants');
        WRITELN;
        FOR I: = 1 TO N DO
            BEGIN
                FOR J: = 1 TO N DO
                    WRITE( A[I,J] :7:4, ' ');
                FOR J : = 1 TO NVEC DO
                    WRITE( ' :', Y[I,J] :7:4);
                WRITELN
            END (* i loop *);
```

*Figure 4.8: Solution of Simultaneous Equations and Matrix Inversion by the
Gauss-Jordan Method (Multiple constant vectors may be entered) (cont.)*

```
            WRITELN
        END  (* if n > 1 *)
END (* procedure get _ data *);

PROCEDURE WRITE _ DATA;
(* print out the answers *)

VAR
    I, J : INTEGER;
BEGIN
    IF NVEC > 0 THEN
        BEGIN
            WRITELN( ' Solution');
            FOR I := 1 TO N DO
                BEGIN
                    FOR J := 1 TO NVEC DO
                        WRITE( Y[I,J] :9:5);
                    WRITELN
                END
        END (* IF *)
    ELSE
        BEGIN
            WRITELN('     Inverse');
            FOR I := 1 TO N DO
                BEGIN
                    FOR J := 1 TO N DO
                        WRITE( A[I,J] :9:5);
                    WRITELN
                END;
            WRITELN;
            WRITE( ' Determinant is ', DETERM)
        END (* ELSE *);
    WRITELN
END (* write _ data *);
```

Figure 4.8: Solution of Simultaneous Equations and Matrix Inversion by the Gauss-Jordan Method (Multiple constant vectors may be entered) (cont.)

```
PROCEDURE GAUSJV
    (VAR B        : ARY2S; (* square matrix of coefficients *)
     VAR W        : ARY2S; (* constant vector matrix *)
     VAR DETERM : REAL;  (* the determinant *)
         NCOL     : INTEGER; (* order of matrix *)
         NV       : INTEGER; (* number of constants *)
     VAR ERROR   : BOOLEAN); (* true if matrix singular *)


(*   Gauss Jordan matrix inversion and solution *)
(*   Nov 21, 80 *)
(*    B(N,N) coefficient matrix, becomes inverse *)
(*    W(N,M) constant vector(s) become solution vector *)
(*    DETERM is determinant *)
(*    ERROR = 1 if singular *)
(*    INDEX(N,3) *)
(*    NV is number of constant vectors *)

LABEL 99;


VAR
    INDEX: ARRAY[1..MAXC, 1..3] OF INTEGER;
    I, J, K, L, IROW, ICOL, N, L1    : INTEGER;
    PIVOT, HOLD, SUM, T, AB, BIG     : REAL;


PROCEDURE SWAP(VAR A, B: REAL);
VAR
    HOLD: REAL;
BEGIN  (* swap *)
    HOLD := A;
    A := B;
    B := HOLD
END (* procedure swap *);
```

Figure 4.8: Solution of Simultaneous Equations and Matrix Inversion by the Gauss-Jordan Method (Multiple constant vectors may be entered) (cont.)

```
PROCEDURE GAUSJ2;

LABEL 98;

VAR
   I, J, K, L, L1: INTEGER;

PROCEDURE GAUSJ3;

VAR
   L: INTEGER;

BEGIN  (* procedure gausj3 *)
   (* interchange rows to put pivot on diagonal *)
   IF IROW <> ICOL THEN
     BEGIN
        DETERM := − DETERM;
        FOR L := 1 TO N DO
           SWAP(B[IROW, L], B[ICOL, L]);
        IF NV > 0 THEN
           FOR L := 1 TO NV DO
           SWAP(W[IROW, L], W[ICOL, L])
     END      (* if irow <> icol *)
END (* gausj3 *);

BEGIN   (* procedure gausj2 *)
   (* actual start of gaussj *)
   ERROR := FALSE;
   N := NCOL;
   FOR I := 1 TO N DO
      INDEX[I, 3] := 0;
   DETERM := 1.0;
```

Figure 4.8: Solution of Simultaneous Equations and Matrix Inversion by the Gauss-Jordan Method (Multiple constant vectors may be entered) (cont.)

```
FOR I := 1 TO N DO
  BEGIN
    (* search for largest element *)
    BIG := 0.0;
    FOR J := 1 TO N DO
      BEGIN
        IF INDEX[J, 3] <> 1 THEN
          BEGIN
            FOR K := 1 TO N DO
              BEGIN
                IF INDEX[K, 3] > 1 THEN
                  BEGIN
                    WRITELN(' ERROR: matrix singular');
                    ERROR := TRUE;
                    GOTO 98 (* abort *)
                  END;
                IF INDEX[K, 3] < 1 THEN
                IF ABS(B[J, K]) > BIG THEN
                  BEGIN
                    IROW := J;
                    ICOL := K;
                    BIG := ABS(B[J, K])
                  END
              END (* k loop *)
          END (* IF *)
      END (* j loop *);
    INDEX[ICOL, 3] := INDEX[ICOL, 3] + 1;
    INDEX[I, 1] := IROW;
    INDEX[I, 2] := ICOL;
    GAUSJ3 (* further subdivision of gaussj *);
    (* divide pivot row by pivot column *)
    PIVOT := B[ICOL, ICOL];
    DETERM := DETERM * PIVOT;
```

Figure 4.8: Solution of Simultaneous Equations and Matrix Inversion by the Gauss-Jordan Method (Multiple constant vectors may be entered) (cont.)

```
            B[ICOL, ICOL] := 1.0;
         FOR L := 1 TO N DO
            B[ICOL, L] := B[ICOL, L]/PIVOT;
         IF NV > 0 THEN
            FOR L := 1 TO NV DO
               W[ICOL, L] := W[ICOL, L]/PIVOT;
         (*  reduce nonpivot rows  *)
         FOR L1 := 1 TO N DO
            BEGIN
               IF L1 <> ICOL THEN
                  BEGIN
                     T := B[L1, ICOL];
                     B[L1, ICOL] := 0;
                     FOR L := 1 TO N DO
                        B[L1, L] := B[L1, L] − B[ICOL, L] * T;
                     IF NV > 0 THEN
                        FOR L := 1 TO NV DO
                           W[L1, L] := W[L1, L] − W[ICOL, L] * T
                  END   (* IF L1 <> icol *)
               END (* FOR L1 *)
         END (* i loop *);
      98:
   END (* gausj2 *);


   BEGIN    (* Gauss—Jordan main program *)
      GAUSJ2 (* first half of gaussj *);
      IF ERROR THEN GOTO 99;
      (* interchange columns *)
      FOR I := 1 TO N DO
         BEGIN
            L := N − I + 1;
```

Figure 4.8: Solution of Simultaneous Equations and Matrix Inversion by the Gauss-Jordan Method (Multiple constant vectors may be entered) (cont.)

```
            IF INDEX[L, 1] <> INDEX[L, 2] THEN
               BEGIN
                    IROW := INDEX[L, 1];
                    ICOL := INDEX[L, 2];
                    FOR K := 1 TO N DO
                         SWAP(B[K, IROW], B[K, ICOL])
                    END  (* if index *)
            END (* i loop *);
        FOR K := 1 TO N DO
          IF INDEX[K, 3] <> 1 THEN
             BEGIN
                 WRITELN(' ERROR: matrix singular');
                 ERROR := TRUE;
                 GOTO 99 (* abort *)
             END;
        99:
    END (* procedure gaussj *);


    BEGIN  (* main program *)
        WRITELN;
        WRITELN(' Simultaneous solution by Gauss—Jordan');
        WRITELN(' Multiple constant vectors, or matrix inverse');
        REPEAT
           GET_DATA (A, Y, N, NVEC);
           IF N > 1 THEN
              BEGIN
                  GAUSJV (A, Y, DETERM, N, NVEC, ERROR);
                  IF NOT ERROR THEN  WRITE_DATA
              END
        UNTIL N < 2
    END.
```

Figure 4.8: Solution of Simultaneous Equations and Matrix Inversion by the Gauss-Jordan Method (Multiple constant vectors may be entered) (cont.)

In the following section we will discuss the Hilbert matrix as an example of ill conditioning. To explore such matrices we will write a program that uses our first version of the Gauss-Jordan Procedure. We will then use the program to solve a series of progressively more problematic Hilbert matrices.

ILL-CONDITIONED EQUATIONS

A singular matrix has a determinant of zero. The corresponding set of equations has either no solution or many solutions. An *ill-conditioned matrix*, by comparison, is one that is *nearly* singular. It produces incorrect or inaccurate answers. A small change in the input data can cause great changes in the answer. A two-dimensional analogue of ill conditioning occurs with two nearly parallel lines. The point of intersection is the desired solution. But, the closer the two lines come to being parallel, the more difficult it is to determine their actual point of intersection.

Various tests for ill conditioning have been suggested. One of these is to compare the values of the inverted matrix to those of the original matrix. If there are differences of several orders of magnitude, then it is likely that ill conditioning is present. Another test is to take the inverse of the inverse of the coefficient matrix and compare the result to the original matrix. They should, of course, be the same. This will test the inversion algorithm and the computer arithmetic at the same time. Yet another test is to compare the magnitudes of values along the major diagonal. They should not be too far apart.

The Hilbert matrix is an example of ill conditioning. This symmetric matrix begins with unity in the upper-left corner. The remaining values get smaller and smaller as we go down a column or across a row, according to the pattern:

$$
\begin{bmatrix}
1 & \frac{1}{2} & \frac{1}{3} & \frac{1}{4} & \cdots\cdots & 1/n \\
\frac{1}{2} & \frac{1}{3} & \frac{1}{4} & \frac{1}{5} & \cdots & 1/(n+1) \\
\frac{1}{3} & \frac{1}{4} & \frac{1}{5} & \frac{1}{6} & \cdots & 1/(n+2) \\
\frac{1}{4} & \frac{1}{5} & \frac{1}{6} & \frac{1}{7} & \cdots & 1/(n+3) \\
\cdots & & & \cdots & & \cdots \\
1/n & \cdots & \cdots & \cdots & \cdots & 1/(2n-1)
\end{bmatrix}
$$

The Hilbert matrix can be used to produce a set of ill-conditioned equations. Consider, for example:

$$x_1 + \frac{x_2}{2} + \frac{x_3}{3} + \frac{x_4}{4} + \frac{x_5}{5} = 1 + \frac{1}{2} + \frac{1}{3} + \frac{1}{4} + \frac{1}{5}$$

$$\frac{x_1}{2} + \frac{x_2}{3} + \frac{x_3}{4} + \frac{x_4}{5} + \frac{x_5}{6} = \frac{1}{2} + \frac{1}{3} + \frac{1}{4} + \frac{1}{5} + \frac{1}{6}$$

$$\frac{x_1}{3} + \frac{x_2}{4} + \frac{x_3}{5} + \frac{x_4}{6} + \frac{x_5}{7} = \frac{1}{3} + \frac{1}{4} + \frac{1}{5} + \frac{1}{6} + \frac{1}{7}$$

$$\frac{x_1}{4} + \frac{x_2}{5} + \frac{x_3}{6} + \frac{x_4}{7} + \frac{x_5}{8} = \frac{1}{4} + \frac{1}{5} + \frac{1}{6} + \frac{1}{7} + \frac{1}{8}$$

$$\frac{x_1}{5} + \frac{x_2}{6} + \frac{x_3}{7} + \frac{x_4}{8} + \frac{x_5}{9} = \frac{1}{5} + \frac{1}{6} + \frac{1}{7} + \frac{1}{8} + \frac{1}{9}$$

First of all, the fractions $\frac{1}{3}$, $\frac{1}{6}$ and $\frac{1}{7}$ cannot be represented exactly. Consequently, there will be roundoff errors at the very beginning of the problem. Secondly, the inverse of the coefficient matrix is exactly:

$$\begin{bmatrix} 25 & -300 & 1050 & -1400 & 630 \\ -300 & 4800 & -18900 & 26880 & -12600 \\ 1050 & -18900 & 79380 & -117600 & 56700 \\ -1400 & 26880 & -117600 & 179200 & -88200 \\ 630 & -12600 & 56700 & -88200 & 44100 \end{bmatrix}$$

Some of the elements of the inverse are orders of magnitude larger than elements of the original matrix. Furthermore, the determinant of the coefficient matrix is nearly zero.

Since each element of the constant vector is the sum of the matrix elements in the corresponding row, the exact solution is:

$$\begin{bmatrix} 1 & 1 & 1 & 1 & 1 \end{bmatrix}$$

But, because of the ill conditioning, the calculated solution might be something like this:

$$\begin{bmatrix} 1.0001 & 0.99816 & 1.00778 & 0.9884 & 1.0056 \end{bmatrix}$$

Now let us look at a program that uses Hilbert matrices to study the effect of ill conditioning.

PASCAL PROGRAM: SOLVING HILBERT MATRICES

The program shown in Figure 4.9 will generate a set of Hilbert matrices and constant vectors corresponding to a solution of

$$\begin{bmatrix} 1 & 1 & 1 & \dots & 1 \end{bmatrix}$$

Running the Program

Type up the program and execute it. It will run automatically. The program begins with the two equations:

$$\begin{bmatrix} 1 & \frac{1}{2} \\ \frac{1}{2} & \frac{1}{3} \end{bmatrix} \quad \begin{bmatrix} \frac{3}{2} \\ \frac{5}{6} \end{bmatrix}$$

which are solved by calling the first version of the Gauss-Jordan procedure. This small matrix is not ill conditioned, and so no problem is apparent. The solution vector is [1 1]. The program then continues with three, four, five, six, and seven equations.

If your Pascal floating-point operations are performed with the usual 32-bit precision, you will start seeing roundoff errors with four equations. Six equations will give accuracy of only two or three significant figures, and the results for seven equations will be meaningless.

Alternately, if you have double-precision, 64-bit floating-point operations, then the situation is very different. In this case, the solution to 17 simultaneous equations will be given to an accuracy of seven significant figures or better. Thus this program can be used to test the significance of your floating-point package.

```
PROGRAM SOLVIT(OUTPUT);
(* Feb  8, 81 *)
(* n-by-n inverse Hilbert matrix *)
(* solution is 1 1 1 1 1 *)

(* Pascal program to perform simultaneous solution *)
(* by Gauss — Jordan elimination *)

CONST
   MAXR = 7;
   MAXC = 7;
```

Figure 4.9: Solution of a Set of Ill-Conditioned Equations

```
TYPE
   ARY   = ARRAY[1..MAXR] OF REAL;
   ARYS  = ARRAY[1..MAXC] OF REAL;
   ARY2S = ARRAY[1..MAXR, 1..MAXC] OF REAL;
VAR
   Y        : ARYS;
   COEF     : ARYS;
   A, B     : ARY2S;
   N, M, I, J: INTEGER;
   ERROR    : BOOLEAN;


PROCEDURE GET _ DATA(VAR A : ARY2S;
                     VAR Y : ARYS;
                     VAR N, M : INTEGER);


(* setup n-by-n Hilbert matrix *)

VAR
   I, J : INTEGER;

BEGIN
   FOR I := 1 TO N DO
     BEGIN
        A[N,I] := 1.0/(N + I - 1);
        A[I,N] := A[N,I]
     END;
   A[N,N] := 1.0/(2 * N - 1);
   FOR I := 1 TO N DO
     BEGIN
        Y[I] := 0.0;
        FOR J := 1 TO N DO
           Y[I] := Y[I] + A[I,J]
     END;
```

Figure 4.9: Solution of a Set of Ill-Conditioned Equations (cont.)

```
        WRITELN;
        IF N < 7 THEN
          BEGIN
            FOR I: = 1 TO N DO
              BEGIN
                FOR J: = 1 TO M DO
                  WRITE( A[I,J] :7:5, ' ');
                  WRITELN( ' : ', Y[I] :7:5)
              END;
            WRITELN
          END  (* if n < 7 *)
END (* procedure get _ data *);

PROCEDURE WRITE _ DATA;
(* print out the answers *)

VAR
   I : INTEGER;

BEGIN
   FOR I : = 1 TO M DO
      WRITE( COEF[I] :9:5);
   WRITELN;
END (* write _ data *);

(* PROCEDURE gaussj
   (VAR a      : ary2s;
    y           : arys;
    VAR coef  : arys;
    ncol        : integer;
    VAR error : boolean);
extern; *)
(*$F GAUSSJ.PAS *)
```

Figure 4.9: Solution of a Set of Ill-Conditioned Equations (cont.)

```
BEGIN  (* main program *)
    A[1,1] := 1.0;
    N := 2;
    M := N;
    REPEAT
        GET _ DATA (A, Y, N, M);
        FOR I := 1 TO N DO
            FOR J := 1 TO N DO
                B[I,J] := A[I,J] (* setup work array *);
        GAUSSJ (B, Y, COEF, N, ERROR);
        IF NOT ERROR THEN WRITE _ DATA;
        N := N + 1;
        M := N
    UNTIL N > MAXR
END.
```

Figure 4.9: Solution of a Set of Ill-Conditioned Equations (cont.)

In the next section we will see another variation of the Gauss-Jordan procedure. We will consider a set of equations in which the number of equations is greater than the number of variables per equation, and we will present a program to compute the "best-fit" solution.

A SIMULTANEOUS BEST FIT

The previous programs in this chapter produce an *exact* fit to a set of linear equations. In each case, the number of unknowns equals the number of independent equations. If, however, there are more unknowns than there are equations, then no unique solution is possible. This situation corresponds to a coefficient matrix having more columns than rows.

There is another case we might consider. Suppose that we want to determine the value of m unknowns by an experimental procedure. If m independent measurements are obtained, then we can find an exact solution. On the other hand, suppose that the number of independent measurements, n, is greater than the number of unknowns, m. Then it is possible to calculate a *best-fit* value for the m unknowns.

As a two-dimensional analogue, consider the three equations:

$$x \; + \; y \; = \; 3$$
$$x \qquad = \; 1$$
$$y \; = \; 1$$

These three lines do not intersect at the same point. Rather, each pair of lines defines a different point on the x-y plane. These points describe a right triangle that has corners at the positions (1, 1), (2, 1), and (1, 2) as shown in Figure 4.10. The best choice for the "intersection" of these three lines is the location of the centroid of the triangle. Since this point is located at one-third of the distance from the base to the apex, then the best fit solution to these three lines is:

$$x \; = \; 1.3333, \;\; y \; = \; 1.3333$$

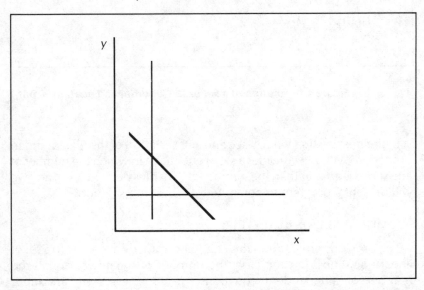

Figure 4.10: Two-Dimensional "Best Fit"

Now that we have illustrated a best-fit solution, we will look at the program that will find the solution.

PASCAL PROGRAM: THE BEST FIT SOLUTION

Make a copy of the program shown in Figure 4.5 and alter it to look like Figure 4.11. Procedure SQUARE, developed in the previous chapter, is included. It is used to convert the rectangular array of coefficients and the constant vector into the square array and vector needed by procedure GAUSSJ. Procedure GAUSSJ is not shown, but is referenced through an INCLUDE directive.

The technique used in this program is essentially the same as that used for *least-squares curve fitting*. Since this topic is covered more fully in the next chapter, it will not be discussed further here.

```
PROGRAM SOLVGJ(INPUT, OUTPUT);
 (* Feb 12, 81 *)
 (* Pascal program to perform simultaneous solution *)
 (* by Gauss—Jordan elimination *)
 (* there may be more equations than unknowns *)


CONST
    MAXR = 8;
    MAXC = 8;
TYPE
    ARY   = ARRAY[1..MAXR] OF REAL;
    ARYS  = ARRAY[1..MAXC] OF REAL;
    ARY2S = ARRAY[1..MAXR, 1..MAXC] OF REAL;
    ARY2  = ARY2S (* for SQUARE *);


VAR
    Y         : ARY;
    COEF, YY : ARYS;
    A, B      : ARY2S;
    N, M, I, J : INTEGER;
    ERROR    : BOOLEAN;


PROCEDURE GET_DATA(VAR A : ARY2S;
                   VAR Y : ARY;
                   VAR N, M : INTEGER);
```

Figure 4.11: The Best Fit to a Set of Linear Equations

```
(* get values for n and arrays a, y *)

VAR
   I, J : INTEGER;

BEGIN
   WRITELN;
   REPEAT
     WRITE( ' How many unknowns? ');
     READLN( M)
   UNTIL M < MAXC;
   IF M > 1 THEN
     BEGIN
       REPEAT
         WRITE( ' How many equations? ');
         READLN( N)
       UNTIL N >= M;
       FOR I := 1 TO N DO
         BEGIN
           WRITELN( ' Equation', I:3);
           FOR J := 1 TO M DO
             BEGIN
               WRITE( J:3, ': ');
               READ( A[I,J])
             END;
           WRITE( ', C: ');
           READLN (Y[I])   (* clear line *)
         END (* i loop *);
       WRITELN;
       FOR I := 1 TO N DO
         BEGIN
```

Figure 4.11: The Best Fit to a Set of Linear Equations (cont.)

```
                    FOR J: = 1 TO M DO
                        WRITE( A[I,J] :7:4, ' ');
                        WRITELN( ' : ', Y[I] :7:4)
                END;
                WRITELN
            END  (* if n > 1 *)
    END (* procedure get_data *);

PROCEDURE WRITE_DATA;
(* print out the answers *)
VAR
    I : INTEGER;

BEGIN
    FOR I := 1 TO M DO WRITE( COEF[I] :9:5);
    WRITELN
END (* write_data *);


(* PROCEDURE square(x : ary2;
                    y : ary;
                VAR a : ary2s;
                VAR g : arys;
            nrow,ncol : integer);
extern; *)
(*$F SQUARE.PAS *)


(* PROCEDURE gaussj
    (VAR b      : ary2s;
     y          : arys;
     VAR coef  : arys;
     ncol       : integer;
     VAR error : boolean);
extern; *)
(*$F GAUSSJ.PAS *)
```

Figure 4.11: The Best Fit to a Set of Linear Equations (cont.)

```
BEGIN (* main program *)
    WRITELN;
    WRITELN(' Best fit to simultaneous equations');
    WRITELN(' By Gauss — Jordan');
    REPEAT
        GET _ DATA (A, Y, N, M);
        IF M > 1 THEN
            BEGIN
                SQUARE(A, Y, B, YY, N, M);
                GAUSSJ (B, YY, COEF, M, ERROR);
                IF NOT ERROR THEN WRITE _ DATA
            END
    UNTIL M < 2
END.
```

Figure 4.11: The Best Fit to a Set of Linear Equations (cont.)

Running the Best-Fit Program

Compile the program and execute it. The number of unknowns is first requested, then the number of equations. If these have the same value, then a regular, simultaneous solution is performed. However, if there are more equations than unknowns, then the best fit is returned. The three equations that we discussed earlier in this section have only two unknowns. The matrix and vector are:

$$\begin{bmatrix} 1 & 1 \\ 1 & 0 \\ 0 & 1 \end{bmatrix} \quad \begin{bmatrix} 3 \\ 1 \\ 1 \end{bmatrix}$$

Verify that the result is:

$$x = 1.3333, \quad y = 1.3333$$

As another example, consider the electric circuit from the beginning of this chapter. We originally derived three loop-current equations that were solved simultaneously. Suppose, however, that the voltages of the sources were determined experimentally. If the value of the left source was found to be 19 volts, and the value of the right source was measured as -5.1 volts, then the three loop equations would be:

$$\begin{bmatrix} 13 & -8 & -3 \\ -8 & 10 & -1 \\ -3 & -1 & 11 \end{bmatrix} \qquad \begin{bmatrix} 19 \\ -5.1 \\ 0 \end{bmatrix}$$

In addition, suppose that the voltage across the horizontal, one-ohm resistor was measured to be 1.1 volts. The current flowing through this resistor is loop current 2 minus loop current 3. Consequently, we can now write an additional independent equation, corresponding to an additional row in our matrix. The four equations are:

$$\begin{bmatrix} 13 & -8 & -3 \\ -8 & 10 & -1 \\ -3 & -1 & 11 \\ 0 & 1 & -1 \end{bmatrix} \qquad \begin{bmatrix} 19 \\ -5.1 \\ 0 \\ 1.1 \end{bmatrix}$$

Enter these four equations, with their three unknowns, into the new program. Compare the resulting best-fit solution:

$$I_1 = 2.8 \text{ amps}, \quad I_2 = 1.8 \text{ amps}, \quad I_3 = 0.94 \text{ amps}$$

to the original solution. This program can be aborted by entering zero for the number of equations.

Next we will devise a method—and a Pascal program—for solving simultaneous equations that have complex coefficients (i.e., factors containing imaginary parts). To illustrate this problem we will study a second, more complicated, electrical example.

EQUATIONS WITH COMPLEX COEFFICIENTS

Simultaneous equations with complex coefficients occur in the analysis of electrical circuits. Complex numbers are not a standard type in Pascal, although they can be defined by the user. In fact, Jensen and Wirth[1] show how to generate complex numbers using record types. Nevertheless, it is rather easy to solve n complex simultaneous equations by converting them into $2n$ equations with real coefficients. The

---

[1]Jensen and Wirth, *Pascal User Manual and Report,* p. 42.

resulting set can then be solved by one of the methods developed previously in this chapter.

Example: An Alternating-Current Electrical Circuit

Consider the electrical circuit shown in Figure 4.12. This circuit is more complicated than Figure 4.1 since it contains an AC power source, an inductor, and a capacitor. The impedance function for a resistor is simply the resistance R. However, the impedance for inductors and capacitors is a function of the frequency. The impedance function for the inductor is $j\omega L$, where j is the imaginary operator equal to the square root of -1, ω is the frequency of the AC source in radians per second, and L is the self inductance in henries. The impedance function for the capacitor is $-j/\omega C$, where C is the capacitance in farads.

For Figure 4.12, the impedance functions are shown next to the corresponding elements. The AC power source is 10 volts (RMS) with a frequency of ω. The inductor has an impedance of $j5$ ohms and the capacitor has an impedance of $-j4$ ohms.

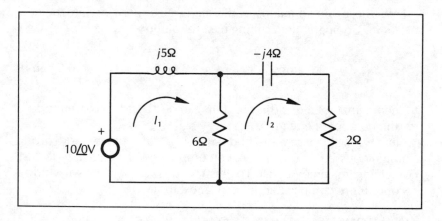

Figure 4.12: A Network Containing an AC Supply

We can find the branch currents for this circuit by using the two loop currents and the Kirchhoff voltage law. A clockwise summing around each loop gives:

$$(6 + j5)I_1 - 6I_2 - 10 \ = \ 0 \quad \text{(left loop)}$$
$$-6I_1 + (8-j4)I_2 \ = \ 0 \quad \text{(right loop)}$$

The corresponding linear equations are:

$$\begin{bmatrix} (6 & + & j5) & (-6 & + & j0) \\ (-6 & + & j0) & (8 & - & j4) \end{bmatrix} \quad \begin{bmatrix} (10 & + & j0) \\ (0 & + & j0) \end{bmatrix}$$

These two equations cannot be directly solved with the programs given previously in this chapter because they contain complex coefficients.

Let us consider a general statement of the two loop equations. The electrical current and the impedance can both be expressed as complex numbers. Consequently, we can write:

$$(AR_{11} + jAI_{11})(IR_1 + jII_1) + (AR_{21} + jAI_{21})(IR_2 + jII_2)$$
$$= (VR_1 + jVI_1)$$

$$(AR_{21} + jAI_{21})(IR_1 + jII_1) + (AR_{22} + jAI_{22})(IR_2 + jII_2)$$
$$= (VR_2 + jVI_2)$$

where the following symbols are used:

AR_{kl} = real part of coefficient (impedance) k,l
AI_{kl} = imaginary part of coefficient k,l
IR_l = real part of current l
II_l = imaginary part of current l
VR_k = real part of voltage for equation k
VI_k = imaginary part of voltage for equation k

Multiplication of the terms on the left of the above equations produces groups that alternately include the complex operator j.

$$(AR_{11} IR_1 - AI_{11} II_1) + j(AR_{11} II_1 + AI_{11} IR_1) + \ldots$$

But if the complex expression on the left is to equal the complex expression on the right, then the real terms on the left must equal the real terms on the right. Similarly, the real coefficients of the imaginary terms on the left must equal the corresponding terms on the right. This approach gives rise to a new set of $2n$ simultaneous equations that contain only real coefficients.

The first new equation is set equal to the real part of the first constant (voltage) term:

$$AR_{11} - AI_{11} + AR_{12} - AI_{12} = VR_1$$

Notice that the complex conjugates of the original coefficients appear in the new first equation. That is, the original coefficients appear in order, but with alternating signs.

The second new equation is set equal to the imaginary part of the first constant (voltage) term:

$$AI_{11} + AR_{11} + AI_{12} + AR_{12} = VI_1$$

This equation also contains all of the coefficients for the first original equation. But in this case, the real and imaginary parts are interchanged. Furthermore, the original signs are utilized.

The complete new equations can be summarized as:

$$\begin{bmatrix} AR_{11} & -AI_{11} & AR_{12} & -AI_{12} \\ AI_{11} & AR_{11} & AI_{12} & AR_{12} \\ AR_{21} & -AI_{21} & AR_{22} & -AI_{22} \\ AI_{21} & AR_{21} & AI_{22} & AR_{22} \end{bmatrix} \begin{bmatrix} VR_1 \\ VI_1 \\ VR_2 \\ VI_2 \end{bmatrix}$$

Substituting the values from Figure 4.12 gives:

$$\begin{bmatrix} 6 & -5 & -6 & 0 \\ 5 & 6 & 0 & -6 \\ -6 & 0 & 8 & 4 \\ 0 & -6 & -4 & 8 \end{bmatrix} \begin{bmatrix} 10 \\ 0 \\ 0 \\ 0 \end{bmatrix}$$

Notice that each original coefficient appears twice in the new matrix. The solution vector for the new set of equations can readily be found by the methods of this chapter. The solution is:

$$\begin{bmatrix} 1.5 & -2.0 & 1.5 & -0.75 \end{bmatrix}$$

which corresponds to the loop currents:

$$I_1 = 1.5 - j2 \quad \text{amps} = 2.5 \; \angle -53°$$
$$I_2 = 1.5 - j0.75 \quad \text{amps} = 1.67 \; \angle -27°$$

These results can be readily verified by calculating the voltages across each circuit element. For example, if the lower node is chosen to be zero volts, then the voltage of the upper node is equal to the voltage across the 6-ohm resistor:

$$V = 6(I_1 - I_2) = 6(-j1.25) = -j7.5 \text{ volts}$$

Similarly, the voltage across the inductor is:

$$V = j5(I_1) = 10 + j7.5 \text{ volts}$$

A sum of the voltages around the left loop then gives:

$$(-j7.5) + (10 + j7.5) - (10) = 0$$

A similar check can be made on the right loop.

Let us now look at a program that will handle these complex coefficients.

PASCAL PROGRAM: SIMULTANEOUS EQUATIONS WITH COMPLEX COEFFICIENTS

The program given in Figure 4.13 simplifies the solution of simultaneous equations with complex coefficients. Each coefficient of the original n equations is entered only once. Then the program converts the data into a $2n$-by-$2n$ matrix and a constant vector of length $2n$. Up to four complex equations can be solved simultaneously. This number can be increased by changing the values of NROW and NCOL. (They must be twice the maximum number of equations.) Any of the methods previously developed in this chapter can find the solution. However, we have selected the Gauss-Jordan technique. Consequently, it might be best to begin with a copy of the Gauss-Jordan program given in Figure 4.5.

```
PROGRAM SOLVEC(INPUT, OUTPUT);
(* Mar 8, 81 *)
(* Pascal program to perform simultaneous solution *)
(* for complex coefficients *)
(* by Gauss — Jordan elimination *)


CONST
    MAXR = 8;
    MAXC = 8;
```

Figure 4.13: Simultaneous Solution of Equations with Complex Coefficients

```
TYPE
  ARY   = ARRAY[1..MAXR] OF REAL;
  ARYS  = ARRAY[1..MAXC] OF REAL;
  ARY2S = ARRAY[1..MAXR, 1..MAXC] OF REAL;
  ARYC2 = ARRAY[1..MAXR, 1..MAXC, 1..2] OF REAL;
  ARYC  = ARRAY[1..MAXR, 1..2] OF REAL;


VAR
  Y       : ARYS;
  COEF    : ARYS;
  A, B    : ARY2S;
  N, M, I, J : INTEGER;
  ERROR   : BOOLEAN;


PROCEDURE GET_DATA(VAR A : ARY2S;
                   VAR Y : ARYS;
                   VAR N, M : INTEGER);

(* get complex values for n and arrays a, y *)

VAR
  C : ARYC2;
  V : ARYC;
  I, J, K, L : INTEGER;

PROCEDURE SHOW;
(* print original data *)

VAR
  I, J, K : INTEGER;

BEGIN (* show *)
  WRITELN;
```

Figure 4.13: Simultaneous Solution of Equations with Complex Coefficients (cont.)

```
      FOR I: = 1 TO N DO
         BEGIN
            FOR J: = 1 TO M DO
               FOR K := 1 TO 2 DO
                  WRITE( C[I,J,K] :7:4, ' ');
               WRITELN( ' : ', V[I,1] :7:4, ' : ', V[I,2] :7:4)
         END;
      N := 2 * N;
      M := N;
      WRITELN;
      FOR I: = 1 TO N DO
         BEGIN
            FOR J: = 1 TO M DO
               WRITE( A[I,J] :7:4, ' ');
            WRITELN( ' : ', Y[I] :9:5)
         END;
      WRITELN
   END (* show *);
BEGIN (* procedure get _ data *)
   WRITELN;
   REPEAT
      WRITE( ' How many equations? ');
      READLN( N);
      M := N
   UNTIL N < MAXR;
   IF N > 1 THEN
      BEGIN
         FOR I := 1 TO N DO
            BEGIN
               WRITELN( 'Equation', I:3);
               K := 0;
               L := 2 * I — 1;
```

Figure 4.13: Simultaneous Solution of Equations with Complex Coefficients (cont.)

```
        FOR J := 1 TO N DO
          BEGIN
            K := K + 1;
            WRITE('Real ', J:3, ': ');
            READ( C[I,J, 1]) (* real part *);
            A[L,K] := C[I,J, 1];
            A[L+1,K+1] := C[I,J, 1];
            K := K + 1;
            WRITE('Imag ', J:3, ': ');
            READ( C[I,J, 2]) (* imaginary part *);
            A[L,K] := −C[I,J, 2];
            A[L+1,K−1] := C[I,J, 2]
          END  (* j loop *);
        WRITE( 'Real const: ');
        READ (V[I,1]) (* real constant *);
        Y[L] := V[I,1];
        WRITE( 'Imag const: ');
        READLN (V[I,2]) (* imag constant *);
        Y[L+1] := V[I,2]
      END (* i loop *);
    SHOW (* original data *)
  END  (* if n > 1 *)
END (* procedure get_data *);


PROCEDURE WRITE_DATA;


(* print out the answers *)


VAR
  I, J : INTEGER;
  RE, IM : REAL;
```

Figure 4.13: Simultaneous Solution of Equations with Complex Coefficients (cont.)

```
FUNCTION MAG(X, Y : REAL): REAL;
(* polar magnitude *)
BEGIN
   MAG := SQRT( SQR(X) + SQR(Y))
END (* function mag *);

FUNCTION ATAN(X, Y : REAL): REAL;
(* arctan in degrees *)
CONST
   PI180 = 57.2957795;

VAR
   A : REAL;

BEGIN (* atan *)
   IF X = 0.0 THEN
      IF Y = 0.0 THEN ATAN := 0.0
      ELSE ATAN := 90.0
   ELSE  (* x <> 0 *)
      IF Y = 0.0 THEN  ATAN := 0.0
      ELSE  (* x and y <> 0 *)
         BEGIN
            A := ARCTAN(ABS(Y/X)) * PI180;
            IF X > 0.0 THEN
               IF Y > 0.0 THEN  ATAN := A (* x, y > 0 *)
               ELSE ATAN := -A (* x > 0, y < 0 *)
            ELSE   (* x < 0 *)
               IF Y > 0.0 THEN  ATAN := 180.0 - A
               (* x < 0, y > 0 *)
               ELSE  ATAN := 180.0 + A (* x, y < 0 *)
         END (* else *)
END (* function atan *);
```

Figure 4.13: Simultaneous Solution of Equations with Complex Coefficients (cont.)

```
BEGIN
   WRITELN
      ('   Real   Imaginary  Magnitude  Angle');
   FOR I := 1 TO M DIV 2 DO
      BEGIN
         J := 2 * I − 1;
         RE := COEF[J];
         IM := COEF[J+1];
         WRITELN( RE :11:5, IM :11:5,
            MAG(RE, IM) :11:5,
            ATAN(RE, IM) :11:5)
      END (* for *);
   WRITELN
END (* write _ data *);


(* PROCEDURE gaussj
   (VAR b    : ary2s;
    y         : arys;
    VAR coef  : arys;
    ncol      : integer;
    VAR error : boolean);
extern; *)
(*$F GAUSSJ.PAS *)


BEGIN  (* main program *)
   WRITELN;
   WRITELN
      (' Simultaneous solution with complex coefficients');
   WRITELN(' By Gauss − Jordan elimination');
   REPEAT
      GET _ DATA (A, Y, N, M);
```

Figure 4.13: Simultaneous Solution of Equations with Complex Coefficients (cont.)

```
        IF N > 1 THEN
            BEGIN
                FOR I := 1 TO N DO
                FOR J := 1 TO N DO
                    B[I,J] := A[I,J] (* setup work array *);
                GAUSSJ (B, Y, COEF, N, ERROR);
                IF NOT ERROR THEN WRITE _ DATA
            END
        UNTIL N < 2
    END.
```

Figure 4.13: Simultaneous Solution of Equations with Complex Coefficients (cont.)

Running the Program

Type up the program and use it to solve the circuit shown in Figure 4.12. Enter the value of 2 for the number of complex equations. Then enter the coefficients for each equation in turn. Give the real coefficient first, then the imaginary part next. The constant vector is entered in the same way; real part first, then imaginary part. The data for this problem are entered as:

$$6, \ 5, \ -6, \ \ \ 0, \ 10, \ 0 \ \text{(equation 1)}$$
$$-6, \ 0, \ \ \ 8, \ -4, \ \ \ 0, \ 0 \ \text{(equation 2)}$$

The original input data are printed out, and the new $2n$-by-$2n$ matrix is given. Then, the solution is given in both rectangular and polar forms.

The polar magnitude is calculated in function MAG by finding the square root of the sum of the squares of the rectangular components. The phasor angle is determined in function ATAN. This function appears to be considerably more complicated than necessary. The problem is that built-in ARCTAN functions have not been standardized. As we noted in Chapter 1, some Pascals return a value in the range of 0° to 180°, and others produce a range from −90° to +90°. For this reason, function ATAN computes the arctan for the first quadrant (0° to 90°). Then the angle is converted to the appropriate quadrant.

We have now seen several methods and variations that are adequate for solving small matrices (i.e., solving small sets of simultaneous equations). The last topic of this chapter will be an iterative method for solving large matrices.

THE GAUSS-SEIDEL ITERATIVE METHOD

The Gauss elimination and Gauss-Jordan methods we considered previously are not suitable for solving very large matrices. More and more multiplication and subtraction operations are performed as the number of equations increases. The resulting roundoff error can produce a meaningless solution.

The Gauss-Seidel method finds the solution to a set of equations by an iterative technique. An initial approximation is repeatedly refined until the result is acceptably close to the solution. Since each approximation depends only on the previous approximation, roundoff error does not accumulate. An added feature is that the equations do not have to be linear.

Consider the three loop-current equations derived from Figure 4.1:

$$
\begin{array}{rcrcrcr}
13I_1 & - & 8I_2 & - & 3I_3 & = & 20 \\
-8I_1 & + & 10I_2 & - & I_3 & = & -5 \\
-3I_1 & - & I_2 & + & 11I_3 & = & 0
\end{array}
$$

We can solve the first equation for I_1:

$$
I_1 = \frac{20 + 8I_2 + 3I_3}{13}
$$

in terms of the other two unknowns. Then if we choose first approximations of zero for I_2 and I_3, we obtain a value of 1.54 for I_1. The second equation is then solved for the second variable:

$$
I_2 = \frac{8I_1 + I_3 - 5}{10}
$$

Substituting the current value of 1.54 for I_1 and zero for I_3 produces a value of 0.73 for I_2. The third equation is similarly solved for the third variable:

$$
I_3 = \frac{3I_1 + I_2}{11}
$$

The current values of 1.54 for I_1 and 0.73 for I_2 give a value of 0.486 for I_3. The process is now repeated. The values of 0.73 for I_2 and 0.486 for I_3 are used to obtain a better value for I_1. After about 20 complete iterations, the values are correct to three significant figures. The following table gives some of the values in the sequence:

I_1	I_2	I_3
0	0	0
1.54	0.73	0.486
2.1	1.23	0.685
2.45	1.53	0.808
2.67	1.71	0.883

There are several potential problems with the Gauss-Seidel method. First of all, the process might not converge. That is, successive values may drift further and further from the correct solution. Consider, for example, the previous three equations. If we write them in reverse order, we will derive the expressions:

$$I_1 = (11 I_3 \quad - \quad I_2) / 3$$
$$I_2 = (13 I_1 \quad - \quad 3 I_3 \quad - \quad 20) / 13$$
$$I_3 = \quad 10 I_2 \quad - \quad 8 I_1$$

First approximations of zero are chosen, as before. However, the subsequent values are clearly diverging:

I_1	I_2	I_3
0	0	0
0	−1.5	−15
−57	−54	−92
−318	−299	−440

The problem in this second case is that the largest values are not located on the major diagonal. The solution is to interchange rows to bring the largest element into the pivot position.

Finally, let us investigate a Pascal implementation of the Gauss-Seidel method. We will make note of a couple of interesting features in this program:

- the differences between *relative* and *absolute criteria* in **IF .. THEN** decisions
- the meaning and use of *point relaxation*.

PASCAL PROGRAM: THE GAUSS-SEIDEL METHOD

Surprisingly, the choice of a first approximation is not too important. An additional matter to be considered, however, is the criterion for convergence. The program shown in Figure 4.14 can be used to explore the Gauss-Seidel method for the solution of linear simultaneous equations. Most of the program can be copied from the Gauss elimination method given in Figure 4.3.

The pivot-interchange routine used in the Gauss elimination program is incorporated here for the same purpose. Rows are interchanged to place the largest element of each column on the major diagonal.

An absolute, rather than a relative criterion is used to determine convergence:

IF ABS(NEXTC — COEF[J]) > TOL **THEN** ...

Normally, it is better to choose a relative criterion:

IF ABS(1 — COEF[J] / NEXTC) > TOL **THEN** ...

But in this case, we must insure that NEXTC will never be zero. One way to do this would be to use the form:

IF ABS(NEXTC — COEF[J]) > ABS(TOL * NEXTC) **THEN** ...

Sometimes the successive approximations jump about wildly. One feature of the program given in Figure 4.14 will reduce this tendency. If two successive approximations differ in sign, then the step size is cut in half.

Another feature of the Gauss-Seidel program shown in Figure 4.14 is known as point relaxation. With this technique, each selected value is a function of the previous iteration, the calculated value, and a relaxation factor, LAMBDA. If COEF[J] is the previous value and NEXTC is the calculated value, then the actual next value becomes:

COEF[J] := LAMBDA * NEXTC + (1.0 — LAMBDA) * COEF[J];

The value of LAMBDA can range from 0 to 2.

```
PROGRAM GAUSID(INPUT, OUTPUT);
(* Feb 23,81 *)
(* Pascal program to perform  *)
(* simultaneous solution by Gauss—Seidel *)
(* procedure SEID is included *)

CONST
  MAXR = 8;
  MAXC = 8;

TYPE
  ARY   = ARRAY[1..MAXR] OF REAL;
  ARYS  = ARRAY[1..MAXC] OF REAL;
  ARY2S = ARRAY[1..MAXR, 1..MAXC] OF REAL;

VAR
  Y       : ARY;
  COEF    : ARYS;
  A       : ARY2S;
  N, M    : INTEGER;
  ERROR   : BOOLEAN;

PROCEDURE GET_DATA
  (VAR A     : ARY2S;
   VAR Y     : ARY;
   VAR N, M  : INTEGER);
(* get values for n and arrays a, y *)

VAR
  I, J : INTEGER;

BEGIN
  WRITELN;
```

Figure 4.14: Solution of Linear Equations by the Gauss-Seidel Method

```
    REPEAT
        WRITE(' How many equations? ');
        READLN(N)
    UNTIL N < MAXR;
    M := N;
    IF N > 1 THEN
        BEGIN
            FOR I := 1 TO N DO
                BEGIN
                    WRITELN(' Equation', I: 3);
                    FOR J := 1 TO N DO
                        BEGIN
                            WRITE(J: 3, ': ');
                            READ(A[I, J])
                        END;
                    WRITE(', C: ');
                    READ(Y[I]);
                    READLN  (* clear line *)
                END;
            WRITELN;
            FOR I := 1 TO N DO
                BEGIN
                    FOR J := 1 TO M DO
                        WRITE(A[I, J]: 7: 4, ' ');
                    WRITELN(' : ', Y[I]: 7: 4)
                END;
            WRITELN
        END  (* if n > 1 *)
    ELSE IF N < 0 THEN N := - N;
    M := N
END (* procedure get _ data *);
```

Figure 4.14: Solution of Linear Equations by the Gauss-Seidel Method (cont.)

```
PROCEDURE WRITE _ DATA;
(* print out the answers *)

VAR
   I : INTEGER;

BEGIN
   FOR I := 1 TO M DO
      WRITE(COEF[I]: 9: 5);
   WRITELN
END (* write _ data *);

PROCEDURE SEID
           (A       : ARY2S;
            Y       : ARY;
       VAR COEF   : ARYS;
            NCOL   : INTEGER;
       VAR ERROR  : BOOLEAN);
(* matrix solution by Gauss—Seidel *)
(* Feb 23, 1981 *)
CONST
   TOL = 1.0E—4;
   MAX = 100;

VAR
   DONE : BOOLEAN;
   I, J, K, L, N : INTEGER;
   NEXTC, HOLD, SUM, LAMBDA, AB, BIG : REAL;

BEGIN
   REPEAT
      WRITE(' Relaxation factor? ');
      READLN(LAMBDA)
```

Figure 4.14: Solution of Linear Equations by the Gauss-Seidel Method (cont.)

```
UNTIL (LAMBDA < 2.0) AND (LAMBDA > 0.0);
ERROR : = FALSE;
N : = NCOL;
FOR I : = 1 TO N − 1 DO
   BEGIN
      BIG : = ABS(A[I, I]);
      L : = I;
      FOR J : = I + 1 TO N DO
         BEGIN
            (* search for largest element *)
            AB : = ABS(A[J, I]);
            IF AB > BIG THEN
               BEGIN
                  BIG : = AB;
                  L : = J
               END
         END (* j loop *);
      IF BIG = 0.0 THEN ERROR : = TRUE
      ELSE
         BEGIN
            IF L <> I THEN
               BEGIN
                  (* interchange rows to put *)
                  (* largest element on diagonal *)
                  FOR J : = 1 TO N DO
                     BEGIN
                        HOLD : = A[L, J];
                        A[L, J] : = A[I, J];
                        A[I, J] : = HOLD
                     END;
                  HOLD : = Y[L];
                  Y[L] : = Y[I];
                  Y[I] : = HOLD
```

Figure 4.14: Solution of Linear Equations by the Gauss-Seidel Method (cont.)

```
                    END      (* if L <> i *)
               END           (* if big *)
          END;               (* i loop *)
     IF A[N, N] = 0.0 THEN ERROR := TRUE
     ELSE
        BEGIN
           FOR I := 1 TO N DO
              COEF[I] := 0.0 (* initial guess *);
           I := 0;
           REPEAT
              I := I + 1;
              DONE := TRUE;
              FOR J := 1 TO N DO
                 BEGIN
                    SUM := Y[J];
                    FOR K := 1 TO N DO
                       IF J <> K THEN
                          SUM := SUM − A[J, K] * COEF[K];
                    NEXTC := SUM/A[J, J];
                    IF ABS(NEXTC − COEF[J]) > TOL THEN
                       BEGIN
                          DONE := FALSE;
                          IF NEXTC * COEF[J] < 0.0 THEN
                             NEXTC := (COEF[J] + NEXTC) * 0.5
                       END;
                    COEF[J] :=
                       LAMBDA * NEXTC + (1.0 − LAMBDA)
                          * COEF[J];
                    WRITELN(I: 4, ', coef(', J, ') =', COEF[J])
                 END  (* J loop *)
           UNTIL DONE OR (I > MAX)
        END; (* IF a[n,n] = 0 *);
```

Figure 4.14: Solution of Linear Equations by the Gauss-Seidel Method (cont.)

```
        IF I > MAX THEN ERROR : = TRUE;
        IF ERROR THEN WRITELN('ERROR: Matrix singular ')
    END (* seid *);


    BEGIN              (* main program *)
        WRITELN;
        WRITELN(' Simultaneous solution by Gauss — Seidel');
        REPEAT
            GET _ DATA(A, Y, N, M);
            IF N > 1 THEN
                BEGIN
                    SEID(A, Y, COEF, N, ERROR);
                    IF NOT ERROR THEN WRITE _ DATA
                END
        UNTIL N < 2
    END.
```

Figure 4.14: Solution of Linear Equations by the Gauss-Seidel Method (cont.)

Running the Gauss-Seidel Program

Generate a copy of Figure 4.14 and execute it. You will be asked for the number of equations. Enter the number 3. Then enter the three equations from the electric circuit shown in Figure 4.1. The order of the equations is now immaterial since we have incorporated a row-interchange routine. You will next be asked for the relaxation factor. Give a value of 1.0. Each successive iteration will be printed out after the iteration number.

Convergence will occur after about 20 iterations. The program will again ask for the number of equations. Give a value of − 3 this time. The minus sign indicates that the equations from the previous step are to be reused.

You can repeatedly rerun the program, trying different values for the relaxation factor. The following table shows the dependence of number of iterations on the choice of the relaxation factor.

Lambda	Iterations
0.8	31
1.0	20
1.2	11
1.3	9
1.4	12
1.5	15
1.8	44

Next, enter the 2-by-2 Hilbert matrix:

$$\begin{bmatrix} 1.0 & 0.5 \\ 0.5 & 0.3333 \end{bmatrix} \quad \begin{bmatrix} 1.5 \\ 0.83333 \end{bmatrix}$$

You will find that the optimum relaxation factor occurs at a value of 1.4, about the same as for the previous set of equations. Yet, for other sets of equations, the optimum value of LAMBDA might be 1.0 or 0.8.

Clearly, the Gauss-Seidel is not as automatic a technique as the others we have considered. But it should be considered for solving large numbers of linear equations, or for solving sets of nonlinear equations.

SUMMARY

We have studied a number of methods for solving simultaneous equations, each method suited to a different situation. We have presented Pascal programs to carry out the algorithms of each of these methods. We have also investigated several special cases: multiple constant vectors; ill-conditioned equations; best-fit solutions for an "overdetermined" equation system; and equations with complex coefficients. In the programs for these special cases we have seen an abundance of new and powerful features of Pascal programming.

CHAPTER **5**

Development of a Curve-Fitting Program

INTRODUCTION

In this chapter we will develop a least-squares curve-fitting program. In particular, we will develop a computer program for finding the best straight line that can represent a set of *x-y* data. This program will generate the data, calculate the desired equation, print out the results, produce a plot of the data, and supply a measure of the correlation between *x* and *y*. A sorting routine will then be added in Chapter 6 to allow handling of real experimental data.

Although the resulting program will be large and complex, we will not program all of the parts at one time. Rather, we will use a modular, top-down approach. Only a small portion of the program will be written at first. This part will be checked by actually running it. Another portion of the code will then be added, and it too will be checked by running the new program. In this way, a relatively large program can be developed in a logical fashion. Each step will be tested along the way. If an error appears, it will most likely be found in the most recently added portion.

As we develop the different parts of this program, we will be discussing a number of algorithms and their implementations. These include:

- the use of a RANDOM function and a "fudge factor" to simulate scattered-line experimental data
- a procedure for plotting graphs on a regular character-oriented terminal or on a line printer
- a least-squares curve-fitting procedure, using differential calculus to arrive at the slope and y-intercept of the actual fitted data
- a simple and elegant method for integrating the correlation coefficient into our program.

THE MAIN PROGRAM

The first thing we will do is write the main program with the input and output routines. The main program will always contain as little as possible: the program name, the declaration statements, and the calls to the various procedures. We will consider two versions of the main program —one that uses a built-in RANDOM function, and another that simulates that function.

First Version: Using the Built-in RANDOM Function

Create the Pascal source program shown in Figure 5.1. Use a descriptive file name such as:

CFIT1.PAS

for the first version. The main program begins with the program name and the declaration of INPUT and OUTPUT. This is immediately followed by the necessary global variables. Then, two procedures are added. One procedure, GET_DATA, provides the input, and the other, WRITE_DATA, performs the output. The calls to these procedures are inserted near the end of the main program. Notice that a semicolon terminates the last procedure call. This is unnecessary, even though it is syntactically correct. The inclusion of the extra semicolon at this time, however, will simplify the programming of the next step.

```
PROGRAM CFIT1(INPUT,OUTPUT);

(* Pascal program to perform a *)
(* linear least—squares fit *)
(* Oct 22, 80 *)

CONST
  MAX = 20;

TYPE
  INDEX = 1..MAX;
  ARY   = ARRAY[INDEX] OF REAL;

VAR
  X, Y, Y_CALC : ARY;
  N     : INTEGER;
  DONE : BOOLEAN;
  A, B   : REAL;

PROCEDURE GET_DATA(VAR X, Y : ARY;
                    VAR N : INTEGER);

(* get values for n and arrays x, y *)
(* y is randomly scattered about a straight line *)

CONST
  A = 2.0;
  B = 5.0;

VAR
  I, J    : INTEGER;
  FUDGE : REAL;
```

Figure 5.1: The Beginning of a Curve-Fitting Program

```
BEGIN
    WRITE( 'Fudge? ');
    READLN( FUDGE);
    IF FUDGE < 0.0 THEN DONE := TRUE
    ELSE
      BEGIN
        REPEAT
          WRITE( 'How many points? ');
          READLN( N)
        UNTIL (N > 2) AND (N <=  MAX);
        FOR I := 1 TO N DO
          BEGIN
            J := N +1 −I;
            X[I] := J;
            Y[I] := (A + B*J)
                    *(1.0 +(2.0*RANDOM(0)−1.0)
                            *FUDGE)
          END   (* FOR loop *)
      END  (* IF *)
END (* procedure get _ data *);




PROCEDURE WRITE _ DATA;

(* print out the answers *)

VAR
    I : INTEGER;
```

Figure 5.1: The Beginning of a Curve-Fitting Program (cont.)

```
BEGIN
   WRITELN;
   WRITELN(' I   X    Y');
   FOR I := 1 TO N DO
      WRITELN(I:3, X[I]:8:1, Y[I]:9:2);
   WRITELN
END (* write _ data *);


BEGIN  (* main program *)
   DONE := FALSE;
   REPEAT
      GET _ DATA (X, Y, N);
      IF NOT DONE THEN
         BEGIN
            WRITE _ DATA;
            (* more lines to be added here *)
         END
   UNTIL DONE
END.
```

Figure 5.1: The Beginning of a Curve-Fitting Program (cont.)

The input procedure GET _ DATA will initially generate a set of *x-y* points using a random number generator. It might be desirable at a later time to alter procedure GET _ DATA so that it will read the experimental data from a disk file.

Second Version: Simulating RANDOM

Procedure GET _ DATA calls a function named RANDOM to generate a randomly scattered straight line. While procedure RANDOM is not formally part of the Pascal language, it may be provided as part of your Pascal package. If your Pascal does not include RANDOM, you will want to use the program shown in Figure 5.2, which contains function RANDOM, rather than the program shown in Figure 5.1.

```
PROGRAM CFIT1A(INPUT,OUTPUT);
(* Pascal program to perform a *)
(* linear least—squares fit *)
(* Oct 22, 80 *)
CONST
   MAX = 20;

TYPE
   INDEX = 1..MAX;
   ARY   = ARRAY[INDEX] OF REAL;

VAR
   X, Y, Y__CALC : ARY;
   N      : INTEGER;
   DONE   : BOOLEAN;
   SEED, A, B : REAL;

FUNCTION RANDOM(DUMMY: INTEGER): REAL;
(* random number 0 — 1 *)
(* define SEED = 4.0 as global *)
CONST
   PI = 3.14159;

VAR
   X: REAL;
   I : INTEGER;

BEGIN  (* Random *)
   X := SEED + PI;
   X := EXP(5.0 * LN(X));
   SEED := X — TRUNC(X);
   RANDOM := SEED
END (* Random *);
```

Figure 5.2: Alternate Routines Including RANDOM

```
PROCEDURE GET_DATA(VAR X, Y : ARY;
                          VAR N : INTEGER);


(* get values for n and arrays x, y *)
(* y is randomly scattered about a straight line *)


CONST
   A = 2.0;
   B = 5.0;


VAR
   I, J    : INTEGER;
   FUDGE : REAL;


BEGIN
   WRITE( 'Fudge? ');
   READLN( FUDGE);
   IF FUDGE < 0.0 THEN DONE := TRUE
   ELSE
      BEGIN
         REPEAT
            WRITE( 'How many points? ');
            READLN( N)
         UNTIL (N > 2) AND (N <=  MAX);
         FOR I := 1 TO N DO
         BEGIN
            J := N +1 −I;
            X[I] := J;
            Y[I] := (A + B*J)
                    *(1.0 +(2.0*RANDOM(0)−1.0)*FUDGE)
         END   (* FOR loop *)
      END  (* IF *)
END; (* procedure get_data *)
```

Figure 5.2: Alternate Routines Including RANDOM (cont.)

```
PROCEDURE WRITE_DATA;
(* print out the answers *)
VAR
   I : INTEGER;
BEGIN
   WRITELN;
   WRITELN(' I   X    Y');
   FOR I := 1 TO N DO
      WRITELN(I:3, X[I]:8:1, Y[I]:9:2);
   WRITELN
END (* write_data *);
BEGIN  (* main program *)
   SEED := 4.0;
   DONE := FALSE;
   REPEAT
      GET_DATA (X, Y, N);
      IF NOT DONE THEN
         BEGIN
            WRITE_DATA;
            (* more lines to be added here *)
         END
   UNTIL DONE
END.
```

Figure 5.2: Alternate Routines Including RANDOM (cont.)

Let us now look more closely at procedure GET_DATA and its algorithm for simulating experimental data.

The Scattering Algorithm

Procedure GET_DATA generates a straight line with an intercept (A) of 2 and a slope (B) of 5. That is, it generates a set of data in the arrays X and Y, corresponding to the line:

$$y = 2 + 5x$$

A random number generator is then used to move the points off the line according to the variable FUDGE. If the value of FUDGE is zero, a perfectly straight line is generated. On the other hand, if FUDGE has a value of 0.2, the points will be displaced to a maximum of 20 percent from the line.

It would be more realistic to generate Gaussian random numbers rather than uniformly distributed random numbers in this application. However, our chosen method is much faster. Furthermore, we will be removing this part of the program when real data are incorporated into the curve-fitting routine.

The scattering algorithm works in the following way. Function RANDOM returns a real number in the range of 0 to 1. This value is doubled to give a range of 0 to 2. The subtraction of 1 sets the range from − 1 to + 1. Finally, this result is multiplied by FUDGE and added to 1 to give the desired range.

Notice that X, Y, and N are passed to procedure GET_DATA through a formal parameter list. This is not really necessary in this case since the procedure could obtain the values globally. This approach, however, effectively isolates the input data from the main program. Suppose, for example, that after reading values for the arrays X and Y, we wanted to make a transformation. We could take the reciprocal of X and the logarithm of Y. Our equation would then be:

$$\ln y = A + \frac{B}{x}$$

The dummy parameter list of GET_DATA could then contain the transformed values. The procedure heading might look like this:

PROCEDURE GET_DATA (**VAR** XR, LOGY : ARY; **VAR** N : INTEGER);

By contrast, procedure WRITE_DATA has no formal parameters.

Running the Main Program

Compile the program and try it out. You will be asked to input a value for FUDGE. Give a value of zero the first time. Then you will be asked for the number of points. Give the value of 9. The result will be three columns of numbers, as shown in Figure 5.3. Notice that the elements of the array X are generated in descending order. We will reorder the arrays X and Y when we add a sorting routine in the next chapter.

At the completion of this task, the program asks for another value for FUDGE. This time, respond with 0.2. The resulting values of x should be the same as the previous values. The values of y, however, will be

```
        I           X           Y
        1          9.0        47.00
        2          8.0        42.00
        3          7.0        37.00
        4          6.0        32.00
        5          5.0        27.00
        6          4.0        22.00
        7          3.0        17.00
        8          2.0        12.00
        9          1.0         7.00
```

Figure 5.3: First Run of the Curve-Fitting Program: FUDGE = 0

somewhat larger or smaller than their previous values. When this step is finished, you will be asked for another value for FUDGE. Enter a negative number this time to terminate the program. Control will then return to the system level.

Now that you have a working program, you can begin to add new features. But always keep copies of the previous versions. Then if you have trouble with a new version, you can return to the previous version and start again.

A PRINTER PLOTTER ROUTINE

Next we will add a routine for plotting the results on an ordinary computer terminal. Large computers are typically provided with a digital plotter and perhaps a graphic video terminal. With these devices, it is possible to display data to a high degree of precision. Unfortunately, this approach is usually time consuming. Furthermore, small computers may not have such devices.

As an alternative, experimental data can be displayed on a regular character-oriented terminal or on a line printer by using plus (+) and star (*) symbols. The resulting plot will be a crude representation of the actual data. However, this plot may be useful for finding gross errors in programming as well as incorrectly entered data. Furthermore, the resulting plot can be viewed immediately along with the computational results, whereas there may be a delay in obtaining plotter output from a large computer.

A Pascal procedure for plotting one or two dependent variables as a function of a third independent variable is shown in Figure 5.4. This routine plots the independent variable vertically rather than in the usual horizontal direction. The dependent variables are then displayed horizontally. That is, the graph is rotated clockwise ¼ turn from the usual axis orientation. We will use this plot to display simple functions. However, multivalued functions can also be displayed. Furthermore, the values of the independent variable need not be uniformly spaced.

```
PROCEDURE PLOT(    (* with arrays *)
                   X, (* as independent variable *)
                   Y, (* as dependent variable *)
                   YCALC (* as fitted curve *) : ARY;
     (* and *) M : INTEGER (* number of pts *));

(* plot y and ycalc as a function of x for m points *)
(* if m is negative, only x and y are plotted *)
(* Feb 9, 81 *)

CONST
    BLANK = ' ';
    LINEL = 51;

VAR
    YLABEL: ARRAY[1..6] OF REAL;
    OUT   : ARRAY[1..LINEL] OF CHAR;
    LINES, I, J, JP, L, N: INTEGER;
    ISKIP, YONLY: BOOLEAN;
    XLOW, XHIGH, XNEXT, XLABEL, XSCALE, SIGNXS,
    YMIN, YMAX, CHANGE, YSCALE, YS10: REAL;

FUNCTION PSCALE(P: REAL): INTEGER;

BEGIN
    PSCALE := TRUNC((P — YMIN)/YSCALE + 1)
END (* pscale *);
```

Figure 5.4: A Plotting Procedure

```
PROCEDURE OUTLIN(XNAME: REAL);
(* output a line *)

VAR
    I, MAX: INTEGER;

BEGIN
    WRITE(XNAME: 8: 2, BLANK) (* line label *);
    MAX := LINEL + 1;
    REPEAT    (* skip blanks on end of line *)
        MAX := MAX - 1
    UNTIL (OUT[MAX] <> BLANK) OR (MAX = 1);
    FOR I := 1 TO MAX DO
        WRITE(OUT[I]);
    WRITELN;
    FOR I := 1 TO MAX DO
        OUT[I] := BLANK    (* blank next line *)
END (* outlin *);

PROCEDURE SETUP(INDEX: INTEGER);
(* set up the plus and asterisk for printing *)

CONST
    STAR = '*';
    PLUS = '+';

VAR
    I: INTEGER;

BEGIN
    I := PSCALE(Y[INDEX]);
    OUT[I] := PLUS;
```

Figure 5.4: A Plotting Procedure (cont.)

```
    IF NOT YONLY THEN
        BEGIN       (* add ycalc too *)
            I := PSCALE(YCALC[INDEX]);
            OUT[I] := STAR
        END
END    (* setup *);


BEGIN       (* body of plot *)
    IF M > 0 THEN   (* plot y and ycalc vs x *)
        BEGIN
            N := M;
            YONLY := FALSE
        END
    ELSE       (* plot only y vs x *)
        BEGIN
            N := − M;
            YONLY := TRUE
        END;
    (* space out alternate lines *)
    LINES := 2 * (N − 1) + 1;
    WRITELN;
    XLOW := X[1];
    XHIGH := X[N];
    YMAX := Y[1];
    YMIN := YMAX;
    XSCALE := (XHIGH − XLOW)/(LINES − 1);
    SIGNXS := 1.0;
    IF XSCALE < 0.0 THEN SIGNXS := −1.0;
    FOR I := 1 TO N DO
        BEGIN
            IF Y[I] < YMIN THEN  YMIN := Y[I];
            IF Y[I] > YMAX THEN  YMAX := Y[I];
```

Figure 5.4: A Plotting Procedure (cont.)

```
            IF NOT YONLY THEN
               BEGIN
                  IF YCALC[I] < YMIN THEN  YMIN := YCALC[I];
                  IF YCALC[I] > YMAX THEN  YMAX := YCALC[I]
               END    (* IF yonly *)
         END;
YSCALE := (YMAX − YMIN)/(LINEL − 1);
YS10 := YSCALE * 10;
YLABEL[1] := YMIN     (* y axis *);
FOR I := 1 TO 4 DO
   YLABEL[I + 1] := YLABEL[I] + YS10;
YLABEL[6] := YMAX;
FOR I := 1 TO LINEL DO
   OUT[I] := BLANK     (* blank line *);
SETUP(1);
L := 1;
XLABEL := XLOW;
ISKIP := FALSE;

FOR I := 2 TO LINES DO     (* set up a line *)
   BEGIN
      XNEXT := XLOW + XSCALE * (I − 1);
      IF ISKIP THEN WRITELN('   −')
      ELSE
         BEGIN
            L := L + 1;
            WHILE
               (X[L] − (XNEXT − 0.5 * XSCALE))*SIGNXS
                  <= 0.0  DO
               BEGIN
                  SETUP(L)  (* setup print line *);
                  L := L + 1
               END (* WHILE *);
```

Figure 5.4: A Plotting Procedure (cont.)

```
                    OUTLIN(XLABEL)   (* print a line *);
                 FOR J := 1 TO LINEL DO
                       OUT[J] := BLANK    (* blank line *)
              END    (* IF iskip *);
          IF (X[L] − (XNEXT + 0.5 * XSCALE))*SIGNXS > 0.0
              THEN ISKIP := TRUE
          ELSE
            BEGIN
                ISKIP := FALSE;
                XLABEL := XNEXT;
                SETUP(L)   (* setup print line *)
            END
        END   (* FOR loop *);
      OUTLIN(XHIGH) (* last line *);
      WRITE(' ');
      FOR I := 1 TO 6 DO
         WRITE('   ^ ');
      WRITELN;
      WRITE(' ');
      FOR I := 1 TO 6 DO
         WRITE(YLABEL[I]: 9: 1, BLANK);
      WRITELN;
      WRITELN
   END   (* plot *);
```

Figure 5.4: A Plotting Procedure (cont.)

Running the Plotter Routine

Make a copy of the previous version of your source program and give it the name CFIT2.PAS. Add procedure PLOT. Place the new procedure just after procedure WRITE _ DATA and just before the start of the **BEGIN** statement of the main program. In addition, you must insert into the main program a call to the new procedure:

PLOT (X, Y, Y, −N);

This function call is placed immediately after the call to procedure WRITE_DATA and just before the **UNTIL** statement. Be sure that there is a semicolon between the procedure calls to WRITE_DATA and to PLOT. The semicolon at the end of the call to procedure PLOT is not needed, but its inclusion now will simplify the next addition to the program. Notice that the second and third arguments in the PLOT call are the same and that the last argument is negative. When the last argument of PLOT is negative, it means that only a single curve is to be plotted.

Compile this version and try it out. Give a value of 0.2 for FUDGE and the plotter output will look like Figure 5.5. On the other hand, if you input a value of zero for FUDGE, you will get a nearly straight line.

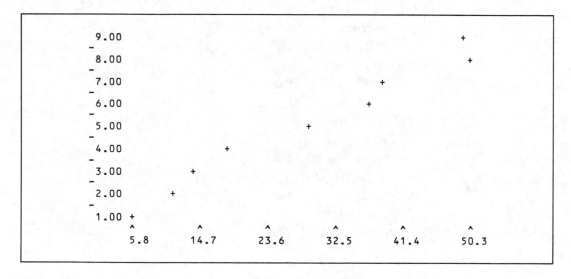

Figure 5.5: y Plotted as a Function of x

We are going to incorporate PLOT into all of the remaining versions of this chapter. Since it is such a large program, you might want to place the procedure into a separate disk file. It can then be referenced by an INCLUDE directive or by an **EXTERN** statement, both of which are described in Chapter 1.

To test all aspects of our plotter routine, we must try plotting two curves at once. Thus, we will incorporate the small procedure LINFIT into our program in the next section. This LINFIT will be replaced by a much more significant procedure later in the chapter, after we have studied the least-squares curve-fitting algorithm.

A SIMULATED CURVE FIT

For our next step, then, we will add a simulated curve-fitting routine. This procedure will simply generate a vector Y_CALC that lies on our straight line with a slope of 5 and an intercept of 2. Keep in mind that we will add a proper curve-fitting routine in a later version.

Add procedure LINFIT, shown in Figure 5.6, to the source program. Place it immediately after procedure WRITE_DATA. Some changes will also have to be made to the main program and to procedure WRITE_DATA. The current version of the main program is shown in Figure 5.8 and the new procedure WRITE_DATA is shown in Figure 5.7. Notice that Y_CALC is now the third parameter in the call to PLOT, and that the final parameter is positive.

```
PROCEDURE LINFIT(X, Y : ARY;
            VAR Y_CALC : ARY;
            VAR A, B  : REAL;
                  N : INTEGER);


(* generate a straight line for x—y *)

VAR
   I : INTEGER;

BEGIN  (* linfit *)
   A := 2.0;
   B := 5.0;
   FOR I := 1 TO N DO
        Y_CALC[I] := A + B * X[I]
END  (* linfit *);
```

Figure 5.6: Procedure LINFIT to Simulate a Linear Fit

```
PROCEDURE WRITE _ DATA;

(* print out the answers *)

VAR
   I : INTEGER;

BEGIN
   WRITELN;
   WRITELN(' I    X    Y    Y CALC');
   FOR I : = 1 TO N DO
      WRITELN(I:3, X[I]:8:1, Y[I]:9:2, Y _ CALC[I]:9:2);
   WRITELN
END (* write _ data *);
```

Figure 5.7: The Revised Procedure WRITE _ DATA

```
BEGIN  (* main program *)
   DONE : = FALSE;
   REPEAT
      GET _ DATA (X, Y, N);
      IF NOT DONE THEN
         BEGIN
            LINFIT(X, Y, Y _ CALC, A, B, N);
            WRITE _ DATA;
            PLOT (X, Y, Y _ CALC, N)
         END
   UNTIL DONE
END.
```

Figure 5.8: The Current Main Program

Running the Plotter Routine with Two Curves

Compile this version and try it out. Again, you will be asked to give a value for FUDGE. Respond with an answer of 0.2. This will produce four columns of data, including the values for Y_CALC. The printer plot that follows the tabular data will show two different curves. Star symbols are used to represent the values of Y_CALC. They will form a nearly perfect straight line. In addition, "+" symbols represent the values of y. They will be scattered on both sides of the Y_CALC values as shown in Figure 5.9. If the two symbols are coincident for a given value of x, only the star is given.

When the program asks for another value of FUDGE, respond with a value of zero. In this case, both y and Y_CALC will have the same values. Since the two lines lie one on top of the other, only one line will be shown. Finally, give a negative value for FUDGE to terminate the program.

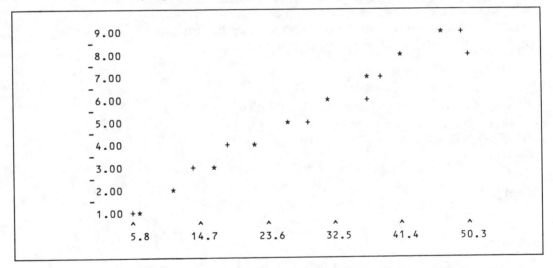

Figure 5.9: y and Y_CALC Plotted Against x

Now that we have written and tested both the main program and a procedure for plotting curves, we are ready to move on to the real topic of this chapter.

THE CURVE-FITTING ALGORITHM

Though it may appear from Figure 5.9 that we have completed our curve-fitting program, we have actually not done so. We have simply generated the values of Y_CALC that correspond to our original line.

The time has come to derive the algorithm for a linear, least-squares procedure. We will first introduce a new vector, **r**, which contains the *residuals*.

For each experimental point, corresponding to an x-y pair, there will be an element of **r** that represents the difference between the corresponding calculated value Y_CALC (which we will represent as $\hat{y}$) and the original value of y. This can be expressed mathematically as:

$$r_i = \hat{y}_i - y_i \tag{1}$$

Occasionally, a point will coincide with the calculated curve. But in general, about half of the x-y points will lie on one side of the fitted curve, resulting in positive values for r. The remaining points will lie on the other side of the curve and give negative values for r. The sum of these residuals should be close to zero.

The least-squares curve-fitting criterion is that the sum of the squares of the residuals be minimized. The square of each residual will be positive, therefore the sum of the residuals squared (*SRS*) will be a positive number. This criterion can be expressed as:

$$SRS = \sum_{i=1}^{n} r_i^2 = \text{ minimum} \tag{2}$$

where n is the number of x-y points (and the length of the vectors **x**, **y**, and $\hat{\mathbf{y}}$).

By combining Equation 1 with the curve-fitting equation:

$$\hat{y}_i = A + Bx_i \tag{3}$$

we get

$$r_i = A + Bx_i - y_i \tag{4}$$

and

$$SRS = \sum_{i=1}^{n} r_i^2 = \sum_{i=1}^{n} (A + Bx_i - y_i)^2 \tag{5}$$

The problem is reduced to finding the values of A and B so that the summation of Equation 5 is minimized. This is accomplished with differential calculus. We take the derivative of Equation 5 with respect to each variable (A and B in this case) and set the result to zero.

$$\frac{\delta \Sigma r_i^2}{\delta A} = 0 \quad \text{and} \quad \frac{\delta \Sigma r_i^2}{\delta B} = 0 \tag{6}$$

Substitution of Equation 5 into Equations 6 gives:

$$\frac{\delta\Sigma (A + Bx_i - y_i)^2}{\delta A} = 0 \tag{7}$$

and

$$\frac{\delta\Sigma (A + Bx_i - y_i)^2}{\delta B} = 0 \tag{8}$$

which is equivalent to:

$$\frac{2\Sigma (A + Bx_i - y_i)\, \delta\Sigma (A + Bx_i - y_i)}{\delta A} = 0 \tag{9}$$

and

$$\frac{2\Sigma (A + Bx_i - y_i)\, \delta\Sigma (A + Bx_i - y_i)}{\delta B} = 0 \tag{10}$$

Since B, x, and y are not functions of A, and the derivative of A with respect to itself is unity, Equation 9 reduces to:

$$\Sigma A + \Sigma Bx_i = \Sigma y_i \tag{11}$$

Similarly, A, x and y are not functions of B. Therefore, Equation 10 becomes:

$$\Sigma Ax_i + \Sigma Bx_i^2 = \Sigma x_i y_i \tag{12}$$

A and B are constants. Therefore they can be factored from the summation step. Equations 7 and 8 can then be expressed as:

$$An + B\Sigma x_i = \Sigma y_i \tag{13}$$

and

$$A\Sigma x_i + B\Sigma x_i^2 = \Sigma x_i y_i \tag{14}$$

We thus have reduced the problem of finding a straight line through a set of x-y data points to one of solving two simultaneous equations (13 and 14). Both these equations are linear in the unknowns A and B. (x, y, and n are, of course, the original data.) The simultaneous solution can be obtained by using Cramer's rule:

$$A = \frac{\begin{vmatrix} \Sigma y_i & \Sigma x_i \\ \Sigma x_i y_i & \Sigma x_i^2 \end{vmatrix}}{\begin{vmatrix} n & \Sigma x_i \\ \Sigma x_i & \Sigma x_i^2 \end{vmatrix}} \tag{15}$$

and

$$B = \frac{\begin{vmatrix} n & \Sigma y_i \\ \Sigma x_i & \Sigma x_i y_i \end{vmatrix}}{\begin{vmatrix} n & \Sigma x_i \\ \Sigma x_i & \Sigma x_i^2 \end{vmatrix}} \tag{16}$$

The corresponding equations we have to solve are:

$$A = \frac{\Sigma x_i^2 \Sigma y_i - \Sigma x_i \Sigma x_i y_i}{n \Sigma x_i^2 - \Sigma x_i \Sigma x_i} \tag{17}$$

and

$$B = \frac{n \Sigma x_i y_i - \Sigma x_i \Sigma y_i}{n \Sigma x_i^2 - \Sigma x_i \Sigma x_i} \tag{18}$$

The computer calculation of A and B is straightforward. The summation of x is obtained by summing the values of the array X. The summation of x^2 is obtained by squaring each value of X and then adding up the squares.

Equations 17 and 18 are commonly converted into an equivalent form by dividing the numerator and denominator by n:

$$A = \frac{(\Sigma x_i^2 \Sigma y_i - \Sigma x_i \Sigma x_i y_i)/n}{\Sigma x_i^2 - \Sigma x_i \Sigma x_i/n} \tag{19}$$

and

$$B = \frac{\Sigma x_i y_i - \Sigma x_i \Sigma y_i/n}{\Sigma x_i^2 - \Sigma x_i \Sigma x_i/n} \tag{20}$$

The denominators of Equations 19 and 20 appear in the formula for the standard deviation discussed in Chapter 2.

We have outlined the mathematics of least-squares curve-fitting and have derived formulas for finding the slope (B) and y-intercept (A) of a linear fitted curve. Now we are ready for the real LINFIT procedure.

The Curve-Fitting Procedure

A procedure for fitting a straight line is shown in Figure 5.10. Replace the original procedure LINFIT with the new one.

```
PROCEDURE LINFIT(X, Y : ARY;
              VAR Y _ CALC : ARY;
              VAR A, B     : REAL;
                    N    : INTEGER);
(* fit a straight line (y_calc) through
     n sets of x and y pairs of points *)
VAR
  I : INTEGER;
  SUM_X, SUM_Y, SUM_XY, SUM_X2,
  SUM_Y2, XI, YI, SXY ,SXX, SYY : REAL;
BEGIN (* linfit *)
  SUM_X := 0.0;
  SUM_Y := 0.0;
  SUM_XY := 0.0;
  SUM_X2 := 0.0;
  SUM_Y2 := 0.0;
  FOR I := 1 TO N DO
    BEGIN
      XI := X[I];
      YI := Y[I];
      SUM_X := SUM_X + XI;
      SUM_Y := SUM_Y + YI;
      SUM_XY := SUM_XY + XI * YI;
      SUM_X2 := SUM_X2 + XI * XI;
      SUM_Y2 := SUM_Y2 + YI * YI
    END;
  SXX := SUM_X2 − SUM_X * SUM_X/ N;
  SXY := SUM_XY − SUM_X * SUM_Y/ N;
  SYY := SUM_Y2 − SUM_Y * SUM_Y/ N;
  B := SXY/SXX;
  A := ((SUM_X2 * SUM_Y − SUM_X * SUM_XY)/ N)/SXX;
  FOR I := 1 TO N DO
    Y_CALC[I] := A + B * X[I]
END (* linfit *);
```

Figure 5.10: Procedure LINFIT to Generate a Least-Squares Fit

At the beginning of procedure LINFIT, the variables SUM_X, SUM_Y, etc., are set to zero and the **FOR** loop is used to calculate the desired sums. Notice that a change of variable is made at the beginning of the loop:

> XI := X[I];
> YI := Y[I];

It generally takes longer to fetch an element of an array than it does to get a scalar value. Consequently, when the same value of an array is needed many times in a loop, it may be faster to define a scalar value and use that instead. On the other hand, some modern compilers incorporate an *optimizer* that automatically performs this task. In this case the transformation is unnecessary.

The arrays X, Y and Y_CALC, and the scalar variables A, B and N are formal parameters as before. With this arrangement, we could call LINFIT to fit X and Y at one point in the program:

> LINFIT (X, Y, Y_CALC, A, B, N);

Then, at a later point, we could again call LINFIT to fit X1 and Y1 with a different number of points:

> LINFIT (X1, Y1, Y1C, A1, B1, M);

For the second call to LINFIT, a straight line with slope B1 and intercept A1 is fitted through the X1, Y1 pairs.

Running the Curve-Fitting Program

Compile the new version and try it out. Give FUDGE a value of 0.2 at first. The fitted line of asterisks should go neatly through the scattered plus symbols. Compare Figure 5.11, which shows the actual fit, with Figure 5.9, which gives the simulated fit. Next, make FUDGE equal to zero. Only a single line of stars should be apparent now. Furthermore, the intercept should be equal to 2 and the slope should be equal to 5, the initial values.

THE CORRELATION COEFFICIENT

Although we now have a procedure for calculating the desired straight line, we are not finished yet. We can obtain the equation of a line fitting our experimental data. Then we can use this equation to

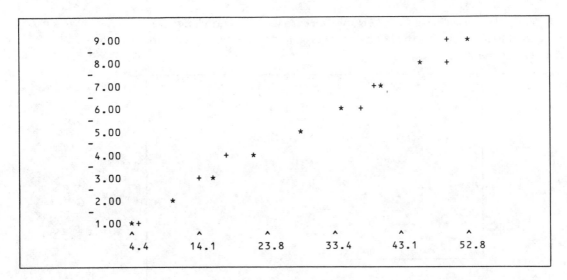

Figure 5.11: A Least-Squares Fit to y vs x

predict a value for y from a given value of x. Under certain circum-
stances, however, our mathematically correct solution is useless.

Consider, for example, the data shown in Figure 5.12. In this case,
procedure LINFIT will find the equation of a straight line through the
data. But, the resulting line does not give us any additional information.
That is, a knowledge of the behavior of x does not tell us anything about
the behavior of y. There is no correlation between x and y.

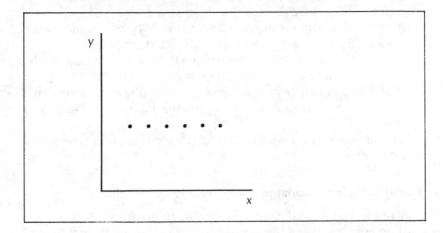

Figure 5.12: No Correlation Between x and y

The data shown in Figure 5.13 is another case where a knowledge of x is no help in predicting the behavior of y. Again, there is no correlation between x and y.

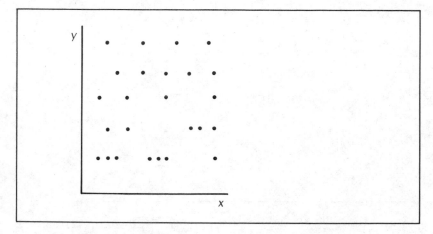

Figure 5.13: Badly Scattered Data

We need to obtain a quantitative measure of the correlation between x and y. We want to know how well we can predict the behavior of y if we know the behavior of x. The measure we need is the *correlation coefficient*.

We saw in Chapter 2 that we could characterize a set of data by the mean and the standard deviation of the values about their own mean. We can also calculate the standard deviation of y about the fitted curve. This measure is termed the *standard error of the estimate (SEE)*. The correlation coefficient compares the standard deviation of y (about its own mean) to the standard deviation about the fitted curve (*SEE*).

The correlation coefficient is zero when there is no correlation. The data in Figures 5.12 and 5.13 are examples of this. On the other hand, the correlation coefficient approaches unity as the data approach a straight line. The correlation coefficient for the data given in Figure 5.11 is 0.99.

We will now incorporate calculations for the correlation coefficient into our program.

Pascal Code for Calculating the Correlation Coefficient

With a few additional statements, our program can calculate the correlation coefficient and the standard errors of the coefficients (the intercept and slope). The standard errors are standard deviations for the coefficients. They can be used to determine confidence intervals for each coefficient.

Add the expression:

CORREL _ COEF, SIGMA _ A, SIGMA _ B, SEE : REAL;

to the **VAR** section at the beginning of the main program. These will then be global variables. The next four lines are placed in procedure LINFIT immediately after the definition for the intercept A:

```
CORREL _ COEF : =
    SXY / SQRT(SXX * SYY);
SEE : =
    SQRT((SUM _Y2 − A * SUM _Y − B * SUM _ XY) / (N − 2));
SIGMA _ B : = SEE / SQRT(SXX);
SIGMA _ A : = SIGMA _ B * SQRT(SUM _ X2 / N);
```

Finally, the output statements in procedure WRITE_ DATA are changed to the following:

```
WRITELN(' Intercept is ', A :8:3,
            ', sigma is ', SIGMA_ A :8:3);
WRITELN('   Slope is ', B :8:2,
            ', sigma is ', SIGMA_ B :8:3);
WRITELN;
WRITELN(' Correlation coefficient is ',
            CORREL_ COEFF :7:4)
```

With the correlation coefficient we have added a final test to our program—a way of quantifying the usefulness of the fitted curve. Let us look now at the finished program.

PASCAL PROGRAM:
LEAST-SQUARES CURVE FITTING FOR SIMULATED DATA

Compile the new version and try it out. First give FUDGE a value of zero. The intercept will be 2 and the slope will be 5, as before. In addition, the sigmas for A and B will be zero and the correlation coefficient will be equal to one. The plot will show a straight line of stars.

Now try a FUDGE of 0.2. This will give sigmas greater than zero for the intercept and the slope. The correlation coefficient will be somewhat less than unity. The entire source program is given in Figure 5.14.

```
PROGRAM CFIT4(INPUT,OUTPUT);
(* Pascal program to perform a *)
(* linear least—squares fit *)
(* Feb 8, 81 *)
CONST
   MAX = 20;
TYPE
   INDEX = 1..MAX;
   ARY   = ARRAY[INDEX] OF REAL;
VAR
   X, Y, Y_CALC : ARY;
   N       : INTEGER;
   DONE  : BOOLEAN;
   A, B, CORREL_COEF,
   SIGMA_A, SIGMA_B, SEE : REAL;


PROCEDURE GET_DATA(VAR X, Y : ARY;
                             VAR N : INTEGER);
(* get values for n and arrays x, y *)
(* y is randomly scattered about a straight line *)
CONST
   A = 2.0;
   B = 5.0;


VAR
   I, J : INTEGER;
   FUDGE : REAL;


BEGIN
   WRITE( 'Fudge? ');
   READLN( FUDGE);
   IF FUDGE < 0.0 THEN DONE := TRUE
   ELSE
```

Figure 5.14: The Complete Curve-Fitting Program

```
    BEGIN
      REPEAT
         WRITE( 'How many points? ');
         READLN( N)
      UNTIL (N > 2) AND (N <= MAX);
      FOR I := 1 TO N DO
         BEGIN
            J := N + 1 - I;
            X[I] := J;
            Y[I] := (A + B*J)
                *(1.0 + (2.0*RANDOM(0) - 1.0)*FUDGE)
         END  (* FOR loop *)
      END (* ELSE *)
END (* procedure get_data *);
PROCEDURE WRITE_DATA;
(* print out the answers *)
VAR
   I : INTEGER;
BEGIN
   WRITELN;
   WRITELN(' I    X     Y   Y CALC');
   FOR I := 1 TO N DO
      WRITELN(I:3, X[I]:8:1,
                Y[I]:9:2, Y_CALC[I]:9:2);
   WRITELN;
   WRITELN(' Intercept is ', A :8:3,
            ', sigma is ', SIGMA_A :8:3);
   WRITELN('    Slope is ', B :8:2,
            ', sigma is ', SIGMA_B :8:3);
   WRITELN;
   WRITELN(' Correlation coefficient is ',
                CORREL_COEFF :7:4)
END (* write_data *);
```

Figure 5.14: The Complete Curve-Fitting Program (cont.)

```
PROCEDURE LINFIT(X, Y : ARY;
            VAR Y_CALC: ARY;
            VAR A, B     : REAL;
                    N  : INTEGER);
(* fit a straight line (y_calc) through
 n sets of x and y pairs of points *)

VAR
  I : INTEGER;
  SUM_X, SUM_Y, SUM_XY, SUM_X2,
  SUM_Y2, XI, YI, SXY ,SXX, SYY : REAL;


BEGIN  (* linfit *)
  SUM_X := 0.0;
  SUM_Y := 0.0;
  SUM_XY := 0.0;
  SUM_X2 := 0.0;
  SUM_Y2 := 0.0;
  FOR I := 1 TO N DO
    BEGIN
      XI := X[I]; YI := Y[I];
      SUM_X := SUM_X + XI;
      SUM_Y := SUM_Y + YI;
      SUM_XY := SUM_XY + XI * YI;
      SUM_X2 := SUM_X2 + XI * XI;
      SUM_Y2 := SUM_Y2 + YI * YI
    END;
  SXX := SUM_X2 − SUM_X * SUM_X / N;
  SXY := SUM_XY − SUM_X * SUM_Y / N;
  SYY := SUM_Y2 − SUM_Y * SUM_Y / N;
  B := SXY/SXX;
  A := ((SUM_X2 * SUM_Y − SUM_X
                * SUM_XY)/ N)/SXX;
```

Figure 5.14: The Complete Curve-Fitting Program (cont.)

```
        CORREL_COEF :=
              SXY/ SQRT( SXX * SYY);
        SEE :=
              SQRT( (SUM_Y2 − A*SUM_Y − B*SUM_XY)
                                      /(N − 2));
        SIGMA_B := SEE/SQRT(SXX);
        SIGMA_A := SIGMA_B * SQRT( SUM_X2/N);
        FOR I := 1 TO N DO
            Y_CALC[I] := A + B * X[I]
    END  (* linfit *);


    (* PROCEDURE plot
              (x, y, y_calc : ary;
                        n : integer);

    extern; *)
    (*$F PLOT.PAS *)


    BEGIN  (* main program *)
      DONE := FALSE;
      REPEAT
        GET_DATA (X, Y, N);
        IF NOT DONE THEN
          BEGIN
            LINFIT(X, Y, Y_CALC, A, B, N);
            WRITE_DATA;
            PLOT (X, Y, Y_CALC, N)
          END
      UNTIL DONE
    END.
```

Figure 5.14: The Complete Curve-Fitting Program (cont.)

SUMMARY

The modular development process that we have used for this program has allowed us to evaluate each of the procedures as we wrote them. We began with the main program and added the procedures step by step to simulate data, plot results, compute the fitted curve, and supply the correlation coefficient. Significantly enough, the curve-fitting procedure, LINFIT, replaced an earlier procedure that we wrote solely for the purpose of testing the plotting routine.

While we now have a linear curve-fitting program that works, it is not very useful. It can only fit data produced by a random number generator. Consequently, we will want to alter procedure GET_ DATA so that it can obtain actual data that is embedded in procedure GET_ DATA, or from the keyboard or a disk file. We will delay this step, however, until we have added a sorting routine in the next chapter.

CHAPTER **6**

Sorting

INTRODUCTION

In this chapter we will develop several different sorting algorithms: two bubble sorts, a Shell sort and both a recursive and a nonrecursive quick sort. Then, we will incorporate one of these routines into the curve-fitting program that we developed in Chapter 5.

First, let us discuss the rationale for including a sort routine in our program.

HANDLING EXPERIMENTAL DATA

The curve-fitting program we wrote in the previous chapter obtained its data from a random number generator. More realistic programs will obtain the data from the keyboard or from a disk file. Alternately, the data can be embedded in the program itself.

In particular, we might want to fit experimental data that have not been acquired in numeric order. As an example, suppose that the thermal expansion of a material is to be investigated. Paired measurements of temperature and the corresponding total length could be taken. The experimental apparatus is brought to a certain temperature and the length is measured. Then the temperature is increased and the sample temperature and length are measured again. However, it would not be wise to continue the experiment in this fashion. The problem is that a length might be measured before the sample temperature became uniform, or before the sample reached the value of the temperature sensor. The resulting measured pairs would all contain errors in the same direction.

A better experimental technique would be to approach the desired temperature sometimes from below and sometimes from above. In this case, the experimental data might look like:

Temperature	Length
100	8.0
300	19.0
200	13.5
500	30.0
400	24.5

The curve-fitting program we developed in the last chapter could readily find a least-squares fit to this data. We could place the data directly in procedure GET _ DATA:

PROCEDURE GET_DATA . . .

     . . .

BEGIN
  N := 5
  X[1] := 100; Y[1] := 8.0;
  X[2] := 300; Y[2] := 19.0;
  X[3] := 200; Y[3] := 13.5;
  X[4] := 500; Y[4] := 30.0;
  X[5] := 400; Y[5] := 24.5
END;

If we now run the program, however, we find that procedure PLOT will give an incorrect rendering of the data. The problem is that the array of independent variables, X, must be arranged either in increasing or in decreasing order. That is, the data must be sorted. Procedure PLOT worked properly in the previous chapter because the X array was generated in decreasing order. Notice that we are not concerned about the ordering of the dependent variable Y.

We will begin our discussion of sorting routines with the easiest one to program—the bubble sort.

A BUBBLE SORT

The process of sorting a collection of items consists of arranging them in increasing or decreasing order. The items might be elements of an array of real numbers, or they might be a group of alphabetic and numeric characters called records. There are many different sorting algorithms. Some are very fast, others are very slow. Some are faster with nearly sorted data and some are slower under these conditions. Some require additional working space, where others utilize only the space occupied by the original data.

The first sorting algorithm we will consider is known as the *bubble sort*. This routine is the easiest to understand and the easiest to program. But, unfortunately, it is also the slowest. For sorting lists that contain fewer than a dozen or so items, however, the difference in speed is unimportant.

With the bubble sort, each element is compared to all of the remaining, unsorted elements. If a particular pair is found to be out of order, the two elements are interchanged. There are two loops, one nested inside the other. The outer loop runs from 1 to one less than the length of the array. The inner loop runs from one larger than the outer loop up to the length of the array. With this algorithm, the smaller items "float" to the top of the array during the sorting process. This is of course the origin of the name "bubble sort".

We will incorporate this first sorting routine into a complete program designed to test the relative efficiency of different sorting routines under different conditions. The rest of the sort routines in this chapter will be listed as procedures, though each one can be incorporated into program TSTSORT.

PASCAL PROGRAM: THE BUBBLE SORT AND TSTSORT

The program shown in Figure 6.1 contains a driver routine in addition to the sorting procedure. When the program is executed, it asks the user for the number of items to be sorted. A random number generator is then called. It generates the desired number of elements in the range of zero to 100. The original set of numbers is printed on the console, 10 numbers per line. The statement CHR(7) sounds the console bell at the beginning and at the end of the sorting process. This will allow a comparison to be made with the other sorting routines to be presented in this chapter. If you are not using an ASCII terminal, then this expression will have to be changed or removed.

At the end of the sorting step, the sorted array is printed on the console. The word *random* is also printed. The console bell sounds a third time, and the sorting procedure is called again. This time, however, the procedure runs on an array that is already sorted. At the end of this second sorting step, the console bell sounds a fourth time and the sorted array is printed out again. The word *sorted* is also printed.

For the third phase of the test program, an array of numbers in reverse order is generated. The sorting routine is then called for the third time. At the end of the step, the sorted array is printed along with the word *reversed*.

Type up the program given in Figure 6.1 and run it. Try to find a length that requires several minutes for sorting. Record the time needed for sorting each of the three arrangements of data. Then a comparison can be made with the other sorting routines given later in this chapter. If you are not using a video terminal, you may want to remove the calls to procedure PRINT so that the arrays will not be printed.

The programs in this chapter require function RANDOM for the generation of random numbers. If your Pascal package does not contain this function, you should include function RANDOM given in Figure 5.2 of Chapter 5.

```
PROGRAM TSTSORT(INPUT, OUTPUT);
(* test speed of sorting routine *)

CONST
    MAX = 1000;
TYPE
    ARY = ARRAY[1..300] OF REAL;
VAR
    X: ARY;
    N, I: INTEGER;
```

Figure 6.1: A Bubble Sort Procedure and Driving Program

```
PROCEDURE PRINT;

VAR
   I: INTEGER;

BEGIN
   WRITELN;
   FOR I := 1 TO N DO
     BEGIN
        WRITE(X[I] :7:2);
        IF (I MOD 10) = 0 THEN WRITELN
     END
END;


PROCEDURE (* bubble *) SORT(VAR A: ARY; N: INTEGER);

VAR
   I,J: INTEGER;
   HOLD: REAL;

BEGIN (* procedure sort *)
   FOR I := 1 TO N − 1 DO
     FOR J := I + 1 TO N DO
        BEGIN
           IF A[I] > A[J] THEN
              BEGIN
                 HOLD := A[I];
                 A[I] := A[J];
                 A[J] := HOLD
              END
        END (* FOR *)
END (* procedure sort *);
```

Figure 6.1: A Bubble Sort Procedure and Driving Program (cont.)

```
BEGIN (* main program *)
  REPEAT
    REPEAT
      WRITELN;
      WRITE(' How many points?');
      READLN(N)
    UNTIL N <= MAX;
    FOR I := 1 TO N DO
      X[I] := 100 * RANDOM(0);
    PRINT;
    WRITE(CHR(7));
    SORT(X, N) (* random numbers *);
    WRITE(CHR(7));
    PRINT;
    WRITELN(' random');
    WRITE(CHR(7));
    SORT(X, N) (* sorted numbers *);
    WRITE(CHR(7));
    PRINT;
    WRITELN(' sorted');
    FOR I := 1 TO N DO
      X[I] := N + 1 - I;
    WRITE(CHR(7));
    SORT(X, N) (* reversed numbers *);
    WRITE(CHR(7));
    PRINT;
    WRITELN(' reversed')
  UNTIL N < 5
END.
```

Figure 6.1: A Bubble Sort Procedure and Driving Program (cont.)

We will now develop a slightly improved bubble sort routine. The reason for this improvement will become clearer as the chapter progresses.

Adding a Separate Swap Routine

Whenever an operation is performed more than once, we should consider using a procedure. Later in this chapter, we will incorporate a sorting procedure into the curve-fitting program from the previous chapter. At that time, we will want to interchange a value of a second array (Y) whenever we interchange the corresponding value of the first array (X). The middle part of the sort routine might then become:

```
BEGIN
    HOLD := A[I];
    A[I] := A[J];
    A[J] := HOLD;
    HOLD := B[I];
    B[I] := B[J];
    B[J] := HOLD
END;
```

In this case, a pair of elements of one array is first interchanged, and then the corresponding elements in the other array are swapped. As in Figure 6.1, the interchange operation requires a third variable called HOLD.

Rather than repeat swap instructions, however, a more elegant approach is to create a swap procedure; we will examine this approach in our second version of the bubble sort routine.

PASCAL PROCEDURE: BUBBLE SORT WITH SWAP

A revised sorting procedure is given in Figure 6.2. The sorting procedure incorporates a separate procedure called SWAP that performs the interchange of two elements. Add this feature to your first sorting program. The source code of this new version is easier to comprehend. Furthermore, the routine can be more easily altered so that two or three arrays, say A, B, and C, can be sorted. The additional lines:

```
SWAP (B[I], B[J]);
SWAP (C[I], C[J]);
```

can be added immediately after the first call to procedure SWAP. The new version will run much faster than the first one when the original array is already ordered. Otherwise, it will run more slowly.

```
PROCEDURE (* bubble *) SORT(VAR A: ARY; N: INTEGER);

(* Adapted from 'Introduction to Pascal,'
        R. Zaks, Sybex, 1980 *)

VAR
   NO_CHANGE: BOOLEAN;
   J: INTEGER;

PROCEDURE SWAP(VAR P, Q: REAL);

VAR
   HOLD: REAL;

BEGIN
   HOLD := P;
   P := Q;
   Q := HOLD
END (* swap *);

BEGIN (* procedure sort *)
   REPEAT
      NO_CHANGE := TRUE;
      FOR J := 1 TO N − 1 DO
         BEGIN
            IF A[J] > A[J+1] THEN
               BEGIN
                  SWAP(A[J], A[J+1]);
                  NO_CHANGE := FALSE
               END
         END (* FOR *)
   UNTIL NO_CHANGE
END (* procedure sort *);
```

Figure 6.2: A Variation of the Bubble Sort: Incorporating a Separate Procedure SWAP

Running the New Version

Incorporate the new version of procedure SORT in place of the original. If you stored the first version of procedure SORT in a separate disk file, make a second copy. Alter the second version so that it looks like the listing in Figure 6.2. Be sure to give the new version of SORT a file name like SORT.PAS or whatever name you use in the INCLUDE directive. Save the original version with a different file name. If procedure SORT is a part of your main program, make a complete duplicate of the program. Then, alter procedure SORT so that it looks like Figure 6.2. Compile and run the new version. The results should be the same as with the first version.

Notice that procedure SWAP is local to procedure SORT. It will not be available to other procedures outside of SORT. If, at a later time, SWAP is needed by other procedures, then it should be made a global procedure. That is, it should be removed from the procedure block of SORT.

Next we will study a more sophisticated and slightly more difficult sort routine.

A SHELL SORT

The major disadvantage with the bubble sort method is that it frequently makes more comparisons and more interchanges than are necessary. An item may be moved from one end of the array to the other. Then, a little later, it might be moved nearly back to where it started.

The *Shell-Metzner* sort is generally more efficient than the bubble sort. Comparisons are initially made over long distances. The first item in the array is compared to one in the middle, rather than to the one right next to it. For short lists of fewer than a dozen items, the Shell sort and bubble sort are comparable in speed. But, as the length of the list increases, the speed of the Shell sort becomes apparent. The Shell sort is noticeably faster than the bubble sort when the number of items exceeds about 50.

We will be able to quantify our comparison of the bubble and Shell sorts by using the TSTSORT program of Figure 6.1.

PASCAL PROCEDURE: THE SHELL-METZNER SORT

Alter your sort routine so that it looks like the listing in Figure 6.3. Notice that procedure SWAP is the same as for the bubble sort. Run the new program and compare the sorting time to the bubble sort. You should find that the Shell sort runs much faster. You will also notice that the Shell sort, like the bubble sort, runs much faster with sorted data than with unsorted data.

```
PROCEDURE (* shell *) SORT(VAR A: ARY; N: INTEGER);

   (* Shell—Metzner sort *)
   (* Adapted from 'Programming in Pascal',
       P. Grogono, Addison—Wesley, 1980 *)

VAR
   DONE: BOOLEAN;
   JUMP, I, J: INTEGER;

PROCEDURE SWAP(VAR P, Q: REAL);

VAR
   HOLD: REAL;

BEGIN
   HOLD := P;
   P := Q;
   Q := HOLD
END (* swap *);

BEGIN
   JUMP := N;
   WHILE JUMP > 1 DO
     BEGIN
       JUMP := JUMP DIV 2;
       REPEAT
         DONE := TRUE;
         FOR J := 1 TO N — JUMP DO
           BEGIN
             I := J + JUMP;
             IF A[J] > A[I] THEN
```

Figure 6.3: A Shell Sort Procedure

```
        BEGIN
            SWAP(A[J], A[I]);
            DONE : = FALSE
        END (* IF *)
    END (* FOR *)
UNTIL DONE
END (* WHILE *)
END (* sort *);
```

Figure 6.3: A Shell Sort Procedure (cont.)

In the discussion of our last and most complex sorting routine we will take up the issue of *recursion*, and we will present recursive and nonrecursive versions of our sort procedure.

THE QUICK SORT

We saw that the bubble sort is easy to program and easy to understand. The Shell sort is a bit more complicated, but it can sort much more quickly. Both of these two algorithms can readily be programmed in other higher-level languages such as BASIC or FORTRAN. The third sorting algorithm we will consider is known as the *quick sort*. It is even more complicated than the previous algorithms. It is generally faster than the bubble sort or the Shell sort, although if the original data are already sorted, or nearly sorted, the Shell sort can be much faster. A quick sort routine takes almost as long to run on a sorted array as on an unsorted one.

PASCAL PROCEDURE: A RECURSIVE QUICK SORT

This first version of quick sort is *recursive*; that is, it contains a procedure that calls itself. A disadvantage of this algorithm is that it cannot be directly converted into BASIC or FORTRAN because of the recursive call.

The bubble sort begins by comparing elements that are side by side. The Shell sort begins by comparing the initial element to one at the middle of the array. Quick sort begins by comparing the elements at the opposite ends of the array. Thus, the initial interchanges for quick sort can be made over large distances.

Procedure PARTIT is repeatedly called to operate on various portions of the array. It divides a given portion into two parts. Then it rearranges the elements so that all of those on the left side are smaller than all of those on the right. The two new sections are then divided into two subsections and the process is repeated. Partitioning continues until there are many sets containing one element each, at which point the array is sorted.

You will want to keep versions of both the Shell sort and quick sort (shown in Figure 6.4). The latter will generally be better. But if you frequently need to sort data that are almost sorted, you may find that Shell sort is more suitable.

```
PROCEDURE (* quick *) SORT(VAR X: ARY; N: INTEGER);
(* a recursive sorting routine *)
(* Adapted from 'The design of Well —
        Structured and Correct Programs,'
        S. Alagic, Springer — Verlag, 1978 *)
PROCEDURE QSORT(VAR X: ARY; M,N: INTEGER);

VAR
   I, J: INTEGER;
PROCEDURE PARTIT(VAR A: ARY; VAR I,J: INTEGER;
                        LEFT, RIGHT: INTEGER);
VAR
   PIVOT: REAL;


PROCEDURE SWAP(VAR P, Q: REAL);
VAR
   HOLD: REAL;


BEGIN
   HOLD := P;
   P := Q;
   Q := HOLD
END (* swap *);
```

Figure 6.4: A Recursive Quick Sort

```
BEGIN
   PIVOT := A[(LEFT + RIGHT) DIV 2];
   I := LEFT;
   J := RIGHT;
   WHILE I <= J DO
      BEGIN
         WHILE A[I] < PIVOT DO
            I := I + 1;
         WHILE PIVOT < A[J] DO
            J := J – 1;
         IF I <= J THEN
            BEGIN
               SWAP(A[I], A[J]);
               I := I + 1;
               J := J – 1
            END
      END (* WHILE *)
END (* partit *);

BEGIN (* qsort *)
   IF M < N THEN
      BEGIN
         PARTIT(X, I, J, M, N) (* divide in two *);
         QSORT(X, M, J) (* sort left part *);
         QSORT(X, I, N) (* sort right part *)
      END
END (* qsort *);

BEGIN (* sort *)
   QSORT(X, 1, N)
END (* sort *);
```

Figure 6.4: A Recursive Quick Sort (cont.)

In the next section we will investigate the subtleties of converting this quick sort into a nonrecursive procedure. A recursive procedure will make a copy of itself and any nonvariable parameters at each level of recursion. However, the array that was sorted with the recursive version of quick sort in Figure 6.4 was declared as a variable parameter in procedures QSORT and PARTIT. Thus, the array was never replicated. For this reason, it is possible to write a nonrecursive version of quick sort. However, the scalar parameters, which are indices for the subarrays, are unique. Consequently, they are duplicated for each recursive call; therefore, it will be necessary to make separate copies of these parameters if we want a nonrecursive version of quick sort.

PASCAL PROCEDURE: A NONRECURSIVE QUICK SORT

Figure 6.5 shows a nonrecursive version of quick sort. The execution time for this nonrecursive version is a little longer than it is for the recursive version. Therefore, it would be better to use the recursive version when programming in Pascal. The nonrecursive version, however, can serve as a template for a FORTRAN or BASIC version, since these languages do not allow recursion. The left and right parameters for the simulated calls are stored in integer arrays LEFT and RIGHT.

The pivot element is initially chosen to be the last element in the array. This is a poor choice for a pivot if the array is already arranged in increasing or decreasing order. Consequently, this pivot element is compared with two other elements: the first element of the array and the element located at the center of the array. The element that is the median of these three, neither the largest nor the smallest, is chosen as the pivot. This change will greatly speed up the sorting process when the array or subarray is already ordered in one direction or the other.

One feature of this version is that at each partitioning, the smaller of the two subsets is sorted before the larger. This minimizes the space needed for storage of the unsorted indices.

Finally, let us return to our curve-fitting program, which is what originally led us into this discussion of sorting routines.

INCORPORATING SORT INTO THE CURVE-FITTING PROGRAM

The next step is to incorporate one of the sorting routines into the curve-fitting program we wrote in Chapter 5. The recursive quick sort routine is probably the best choice, but you may want to choose the Shell sort instead. As before, keep a working copy of the prior version. Then you will have something to go back to if the new version becomes hopelessly mixed up. Two changes will have to be made to the sorting

```
PROCEDURE (* quick *) SORT(VAR X: ARY; N: INTEGER);


(* a nonrecursive quicksort routine *)
(* Adapted from 'Software Tools',
      B. Kernighan, Addison—Wesley, 1976 *)


VAR
   LEFT, RIGHT: ARRAY[1..20] OF INTEGER;
   I, J, SP,MID: INTEGER;
   PIVOT: REAL;


PROCEDURE SWAP( VAR P, Q: REAL);


VAR
   HOLD: REAL;


BEGIN
   HOLD := P;
   P := Q;
   Q := HOLD
END;


BEGIN
   LEFT[1] := 1;
   RIGHT[1] := N;
   SP := 1;
   WHILE SP > 0 DO
      BEGIN
         IF LEFT[SP] >= RIGHT[SP] THEN SP := SP — 1
         ELSE
```

Figure 6.5: A Nonrecursive Version of Quick Sort

```
BEGIN
   I := LEFT[SP];
   J := RIGHT[SP];
   PIVOT := X[J];
   MID := (I + J) DIV 2;
   IF (J − I) > 5 THEN
      IF  ((X[MID] < PIVOT) AND (X[MID] > X[I]))
        OR
        ((X[MID] > PIVOT) AND (X[MID] < X[I]))
           THEN SWAP( X[MID], X[J])
      ELSE
         IF ((X[I] < X[MID]) AND (X[I] > PIVOT))
           OR ((X[I] > X[MID]) AND (X[I] < PIVOT))
              THEN SWAP( X[I], X[J]);
   PIVOT := X[J];
   WHILE  I < J DO
      BEGIN
         WHILE X[I] < PIVOT DO
            I := I + 1;
         J := J − 1;
         WHILE (I < J) AND (PIVOT < X[J]) DO
            J := J − 1;
         IF I < J THEN SWAP(X[I], X[J])
      END (* WHILE *);
   J := RIGHT[SP] (* pivot to i *);
   SWAP(X[I], X[J]);
   IF I − LEFT[SP] >= RIGHT[SP] − I THEN
      BEGIN (* put shorter part first *)
         LEFT[SP+1] := LEFT[SP];
         RIGHT[SP+1] := I − 1;
         LEFT[SP] := I + 1
      END
```

Figure 6.5: A Nonrecursive Version of Quick Sort (cont.)

```
              ELSE
                  BEGIN
                      LEFT[SP+1] := I + 1;
                      RIGHT[SP+1] := RIGHT[SP];
                      RIGHT[SP] := I - 1
                  END;
                  SP := SP + 1 (* push stack *)
              END (* IF *)
          END (* WHILE *)
      END (* quick sort *);
```

Figure 6.5: A Nonrecursive Version of Quick Sort (cont.)

routine so that the array Y is sorted along with array X. One change is to add the array Y to the parameter list in the heading. The other change is to add another call to procedure SWAP.

If you choose quick sort, change the heading to read:

PROCEDURE (* QUICK *) SORT
(**VAR** X, Y: ARY;
N: INTEGER);

Then, near the end of procedure PARTIT, add the line:

SWAP(Y[I], Y[J]);

immediately after the similar call to swap:

SWAP(A[I], A[J]);

The sorting routine can be added to the linear fit program in one of several ways. The simplest way is to place the entire procedure immediately after procedure GET_DATA. Then, place the statement:

SORT (X, Y, N);

in the main program. If the sort routine is called after the linear fit is calculated, we will have to operate on three arrays: X, Y, and Y_CALC. Therefore, we will want to perform the sorting step immediately after obtaining the data from procedure GET_DATA. Consequently, place the call to procedure SORT right after the command:

GET _ DATA (X, Y, N);

near the end of the main program.

Using INCLUDE or EXTERN for the Sort Procedure

As an alternate approach, you might want to place the sort procedure in a separate disk file called SORT.PAS and refer to it with an include directive:

(* $I SORT.PAS *)

Place this line just before the directive for PLOT. This latter approach is much more versatile. You can easily try out each of the four sorting routines without editing the main curve-fitting program. Of course, it may be necessary to recompile the combination. Suppose that the bubble sort, the Shell sort and the quick sort routines have the following file names:

SORTB.PAS
SORTS.PAS
SORTQ.PAS
SORTN.PAS

Then you can rename each one in turn to be SORT.PAS and rerun the main program.

The third possible arrangement is to place the sorting routines into a separate disk file and refer to them with an external statement:

PROCEDURE SORT(**VAR** X, Y: ARY;
                             N: INTEGER);
    EXTERN;

Rerun the curve-fitting program after the sort routine has been incorporated. Verify that the elements of array X are sorted in increasing order. Check that the array Y has been rearranged along with X. The slope of the data displayed on the plot should now be positive.

SUMMARY

We have seen variations of three common sort routines: the bubble sort, the Shell sort and the quick sort. All of these sorting routines are written to operate on real numbers. Keep in mind that these procedures can be easily altered to sort integers, characters, or strings of characters. The only caution is that the variable HOLD must be declared to be the same type as the elements of the array X being sorted.

Now that our curve-fitting program can handle real experimental data, we are ready to apply the program to some more complex equations, which is the topic of Chapter 7.

CHAPTER 7

General Least-Squares Curve Fitting

INTRODUCTION

In this chapter we will develop several different least-squares curve-fitting programs. Our goal will be to *generalize* the previous curve-fitting program so that it will be a more useful and realistic tool for a wider variety of experimental situations. Up to now we have been limited to the equation of a straight line. Our first program in this chapter will implement a parabolic curve—that is, a second-order polynomial equation. From that point we will develop a method that will handle both higher-order polynomials and nonpolynomial equations. The only restriction will be that the unknown coefficients must be linear.

The key to our approach will be the use of a data vector and a data matrix. The method will subsequently allow us to input the *order* of a polynomial equation from the keyboard once the program is running. Finally, we will implement curve-fitting programs for some real experimental data. The equations of the fitted data will include those used for heat-capacity and vapor pressure. In addition, in our program for a three-variable equation of state, we will experiment with solutions for a nonlinear coefficient.

A PARABOLIC CURVE FIT

A least-squares curve-fitting program was developed in Chapter 5. This program was used to calculate the coefficients A and B for the expression:

$$y = A + Bx \tag{1}$$

While Equation 1 is the most commonly used curve-fitting equation, there are times when a different equation is required. Therefore, in this chapter, we will extend the least-squares method to include other commonly used expressions.

We will place one restriction on the form of the curve-fitting equation. It must be linear in the unknown coefficients. Thus, we can consider equations such as:

$$y = A + Bx + Cx^2$$

$$y = A + \frac{B}{x} + Cz$$

$$\ln y = \frac{Ax}{z} + B\,e^x$$

because the coefficients A, B, and C are linear. But we will not consider an equation such as:

$$y = A + B\,e^{Cx}$$

since the coefficient C is not linear.

The development of a curve-fitting program for any of the above equations is similar to the approach we used in Chapter 5. Consider, for example, the parabolic equation, which is a second-order polynomial:

$$y = A + Bx + Cx^2 \tag{2}$$

We define the residuals to be:

$$r = A + Bx + Cx^2 - y$$

The residuals are first squared and then summed. As in Chapter 5, we wish to minimize this quantity. We thus take the derivative with respect to each variable (A, B, and C in this case), and set the resulting equations to zero. For the parabolic curve fit there will be three equations, one for each of the three variables. The resulting equations are:

$$An + B\Sigma x + C\Sigma x^2 = \Sigma y \tag{3}$$

$$A\Sigma x + B\Sigma x^2 + C\Sigma x^3 = \Sigma xy \tag{4}$$

$$A\Sigma x^2 + B\Sigma x^3 + C\Sigma x^4 = \Sigma x^2 y \tag{5}$$

PASCAL PROGRAM:
LEAST-SQUARES CURVE FIT FOR A PARABOLA

The solution to the parabolic fit is obtained by solving Equations 3, 4, and 5 simultaneously. The program shown in Figure 7.1 finds the solution to these equations by using Cramer's rule. The determinants of four 3-by-3 matrices must be solved with this approach. Function DETER, developed in Chapter 4 (Figure 4.2), is used for this purpose.

```
PROGRAM LEAST1(OUTPUT);
   (* Dec 26, 80 *)

(* Pascal program to perform a *)
(* linear least—squares fit *)
(* using a parabolic curve *)
(* separate procedure PLOT needed *)

CONST
   MAXR = 20;
   MAXC = 3;

TYPE
   ARY   = ARRAY[1..MAXR] OF REAL;
   ARYS  = ARRAY[1..MAXC] OF REAL;
   ARY2S = ARRAY[1..MAXC, 1..MAXC] OF REAL;
```

Figure 7.1: A Parabolic Least-Squares Fit

```
VAR
   X, Y, Y_CALC : ARY;
   COEF          : ARYS;
   NROW, NCOL  : INTEGER;
   CORREL_COEF: REAL;

PROCEDURE GET_DATA(VAR X,  Y   : ARY;
                   VAR NROW : INTEGER);

(* get values for nrow and arrays x, y *)

VAR
   I: INTEGER;

BEGIN
   NROW := 9;
   FOR I := 1 TO NROW DO X[I] := I;
   Y[1] := 2.07; Y[2] := 8.6;
   Y[3] := 14.42; Y[4] := 15.80;
   Y[5] := 18.92; Y[6] := 17.96;
   Y[7] := 12.98; Y[8] := 6.45;
   Y[9] := 0.27
END (* procedure get_data *);


PROCEDURE WRITE_DATA;

(* print out the answers *)

VAR
   I: INTEGER;
```

Figure 7.1: A Parabolic Least-Squares Fit (cont.)

```
BEGIN
  WRITELN;
  WRITELN('      I      X       Y        Y CALC');
  FOR I := 1 TO NROW DO
    WRITELN(I:3, X[I]:8:1, Y[I]:9:2, Y_CALC[I]:9:2);
  WRITELN; WRITELN(' Coefficients');
  FOR I := 1 TO NCOL DO
    WRITELN( COEF[I]:8:4);
  WRITELN;
  WRITELN(' Correlation coefficient is ',
            CORREL_COEF:8:5)
END (* write_data *);


PROCEDURE SOLVE( A : ARY2S;
                 Y : ARYS;
             VAR COEF : ARYS;
                 NROW : INTEGER;
             VAR ERROR : BOOLEAN);


VAR
  B    : ARY2S;
  I, J : INTEGER;
  DET : REAL;


FUNCTION DETER(A: ARY2S): REAL;
(* calculate the determinant of a 3—by—3 matrix *)

BEGIN (* function deter *)
  DETER := A[1,1] * (A[2,2] * A[3,3] — A[3,2] * A[2,3])
         — A[1,2] * (A[2,1] * A[3,3] — A[3,1] * A[2,3])
         + A[1,3] * (A[2,1] * A[3,2] — A[3,1] * A[2,2])
END (* function deter *);
```

Figure 7.1: A Parabolic Least-Squares Fit (cont.)

```
PROCEDURE SETUP(VAR B      : ARY2S;
                     VAR COEF : ARYS;
                            J    : INTEGER);
VAR
   I: INTEGER;

BEGIN (* setup *)
   FOR I := 1 TO NROW DO
      BEGIN
          B[I,J] := Y[I];
          IF J > 1 THEN B[I,J−1] := A[I,J−1]
      END;
   COEF[J] := DETER(B)/DET
END (* setup *);


BEGIN (* procedure solve *)
   ERROR := FALSE;
   FOR I := 1 TO NROW DO
      FOR J := 1 TO NROW DO
         B[I,J] := A[I,J];
   DET := DETER(B);
   IF DET = 0.0 THEN
      BEGIN
         ERROR := TRUE;
         WRITELN(' ERROR: matrix singular')
      END
   ELSE
      BEGIN
         SETUP( B, COEF, 1);
         SETUP( B, COEF, 2);
         SETUP( B, COEF, 3)
      END (* else *)
END (* procedure solve *);
```

Figure 7.1: A Parabolic Least-Squares Fit (cont.)

```
PROCEDURE LINFIT(X, Y: ARY;
          VAR Y _ CALC  : ARY;
          VAR COEF      : ARYS;
              NROW      : INTEGER;
          VAR NCOL      : INTEGER);


(* least squares fit to a parabola *)
(* nrow sets of x and y pairs of points *)

VAR
    A : ARY2S;
    G : ARYS;
    I : INTEGER;
    ERROR: BOOLEAN;
    SUM _ X, SUM _ Y, SUM _ XY, SUM _ X2,
    SUM _ Y2, XI, YI, SXY ,SXX, SYY,
    SUM _ X3, SUM _ X4, SUM _ 2Y, DENOM,
    SRS, X2: REAL;

BEGIN (* linfit *)
    NCOL := 3 (* polynomial terms *);
    SUM _ X := 0;
    SUM _ Y := 0;
    SUM _ XY := 0;
    SUM _ X2 := 0;
    SUM _ Y2 := 0;
    SUM _ X3 := 0;
    SUM _ X4 := 0;
    SUM _ 2Y := 0;
    FOR I := 1 TO NROW DO
```

Figure 7.1: A Parabolic Least-Squares Fit (cont.)

```
      BEGIN
        XI := X[I];
        YI := Y[I];
        X2 := XI * XI;
        SUM_X := SUM_X + XI;
        SUM_Y := SUM_Y + YI;
        SUM_XY := SUM_XY + XI * YI;
        SUM_X2 := SUM_X2 + X2;
        SUM_Y2 := SUM_Y2 + YI * YI;
        SUM_X3 := SUM_X3 + XI * X2;
        SUM_X4 := SUM_X4 + X2 * X2;
        SUM_2Y := SUM_2Y + X2 * YI
      END;
   A[1,1] := NROW;
   A[2,1] := SUM_X; A[1,2] := SUM_X;
   A[3,1] := SUM_X2; A[1,3] := SUM_X2;
   A[2,2] := SUM_X2; A[3,2] := SUM_X3;
   A[2,3] := SUM_X3; A[3,3] := SUM_X4;
   G[1] := SUM_Y;
   G[2] := SUM_XY;
   G[3] := SUM_2Y;
   SOLVE( A, G, COEF, NCOL, ERROR);
   SRS := 0.0;
   FOR I := 1 TO NROW DO
     BEGIN
        Y_CALC[I] :=
          COEF[1] + COEF[2] * X[I] + COEF[3] * SQR(X[I]);
        SRS := SRS + SQR(Y[I] − Y_CALC[I])
     END;
   CORREL_COEF :=
   SQRT(1.0 − SRS/(SUM_Y2 −SQR(SUM_Y)/NROW))
END (* linfit *);
```

Figure 7.1: A Parabolic Least-Squares Fit (cont.)

```
(*
PROCEDURE plot
          (x, y, y _ calc: ary; nrow: integer);
extern; *)
(*$F PLOT.PAS *) (* get procedure PLOT *)
BEGIN (* main program *)
   GET _ DATA(X, Y, NROW);
   LINFIT(X, Y, Y _ CALC, COEF, NROW, NCOL);
   WRITE _ DATA;
   PLOT(X, Y, Y _ CALC, NROW)
END.
```

Figure 7.1: A Parabolic Least-Squares Fit (cont.)

In the programs of the previous chapters, the coefficients were represented by A, B, and C. However, in this chapter, we will use the vector COEF for this purpose. Thus, the constant term, *A*, now corresponds to the first element of the vector, COEF[1]. Similarly, the coefficients *B* and *C* correspond to COEF[2] and COEF[3], respectively.

Notice that procedure PLOT, which we used in previous chapters, is called by the main program. The plot routine, however, is not shown in the listing. Rather, it is referenced through an INCLUDE directive. If your Pascal compiler does not have this feature, then you will have to insert a copy of procedure PLOT into the program.

Procedure PLOT requires the independent-variable array to be arranged in increasing or decreasing order. The data provided by procedure GET _ DATA are so arranged. However, at a later time, you may want to substitute other data that are not sorted. In this case, you can include one of the sorting procedures developed in Chapter 6.

Running the Program

Type up this program and execute it. The results should look like Figure 7.2. The correlation coefficient is close to unity, indicating that the parabolic equation can produce a good fit to the data. The resulting equation is:

$$y = -7.827 + 10.59x - 1.083x^2$$

Now let us consider polynomial equations with orders higher than 2, and nonpolynomial equations. We will see that the approach we have been using to solve for the coefficients becomes less practical as the equation becomes more complex. Therefore, we will want to investigate another method.

CURVE FITS FOR OTHER EQUATIONS

If a higher-order polynomial is chosen, there will be additional coefficients, resulting in additional equations to be solved simultaneously.

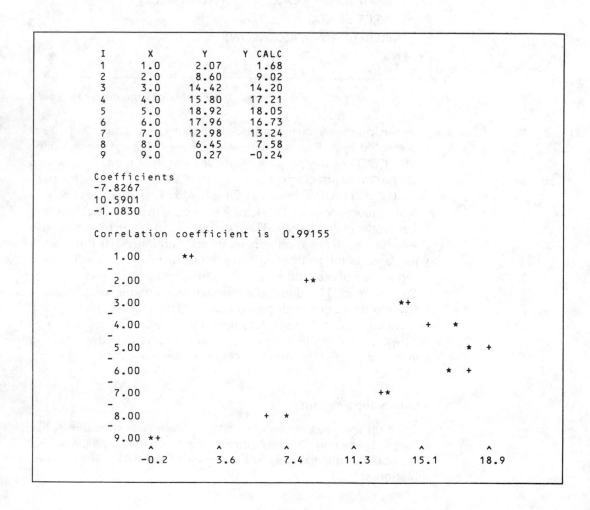

```
   I      X        Y      Y CALC
   1     1.0     2.07     1.68
   2     2.0     8.60     9.02
   3     3.0    14.42    14.20
   4     4.0    15.80    17.21
   5     5.0    18.92    18.05
   6     6.0    17.96    16.73
   7     7.0    12.98    13.24
   8     8.0     6.45     7.58
   9     9.0     0.27    -0.24

Coefficients
-7.8267
10.5901
-1.0830

Correlation coefficient is  0.99155

   1.00      *+
     -
   2.00                      +*
     -
   3.00                              *+
     -
   4.00                                 +    *
     -
   5.00                                      *   +
     -
   6.00                                    *   +
     -
   7.00                          +*
     -
   8.00                 +    *
     -
   9.00  *+
        ^         ^         ^         ^         ^         ^
       -0.2      3.6       7.4      11.3      15.1      18.9
```

Figure 7.2: Output from the Parabolic Curve Fit Program

Equations 3, 4, and 5 can be easily extended in this case. For example, to find the coefficients of the cubic equation:

$$y = A + Bx + Cx^2 + Dx^3$$

the residuals are defined as:

$$r = A + Bx + Cx^2 + Dx^3 - y$$

The residuals are first squared and then summed. The derivative is taken with respect to each of the four variables A, B, C, and D. The resulting four equations are solved simultaneously.

$$An + B\Sigma x + C\Sigma x^2 + D\Sigma x^3 = \Sigma y$$
$$A\Sigma x + B\Sigma x^2 + C\Sigma x^3 + D\Sigma x^4 = \Sigma xy$$
$$A\Sigma x^2 + B\Sigma x^3 + C\Sigma x^4 + D\Sigma x^5 = \Sigma x^2 y$$
$$A\Sigma x^3 + B\Sigma x^4 + C\Sigma x^5 + D\Sigma x^6 = \Sigma x^3 y$$

Nonpolynomial expressions are treated similarly. For example, the general three-term equation is:

$$y = A + Bf(x) + Cg(x)$$

where $f(x)$ and $g(x)$ represent any function of x. The residuals, defined as:

$$r = A + B f(x) + C g(x) - y$$

are squared and then summed. The derivatives with respect to the three variables give the three equations:

$$An + B\Sigma f(x) + C\Sigma g(x) = \Sigma y$$
$$A\Sigma f(x) + B\Sigma f(x)^2 + C\Sigma f(x)g(x) = \Sigma f(x)y$$
$$A\Sigma g(x) + B\Sigma f(x)g(x) + C\Sigma g(x)^2 = \Sigma g(x)y$$

While the above approach to curve fitting is correct, it is laborious. Major alterations are needed whenever the curve-fitting equation is changed. For example, if the equation were chosen to be:

$$y = A + Bx + \frac{C}{x^2}$$

then there would need to be statements in the program such as:

$$SUM\_X3 := SUM\_X3 + 1/XI;$$
$$SUM\_X4 := SUM\_X4 + 1/(XI * XI);$$

for calculating the needed sums.

A Direct Solution

A better way to determine the coefficients of the curve-fitting equation is to set up the data in a matrix and a vector. The data matrix and data vector are then converted to a set of simultaneous equations that are solved in the usual way. For example, suppose that we want a linear fit to the equation

$$y = A + Bx$$

for five sets of x-y data. The data vector, in this case, would simply be the vector of y values. The data matrix would look like this:

$$\begin{bmatrix} 1 & x_1 \\ 1 & x_2 \\ 1 & x_3 \\ 1 & x_4 \\ 1 & x_5 \end{bmatrix}$$

Each row of the data matrix corresponds to one of the data points, whereas each column of the matrix corresponds to one term of the equation. Consequently, the data matrix has 5 rows and 2 columns. Column 1 of the data matrix contains only the value of 1, since that is the corresponding function of x (i.e., x^0) in the first term of the equation. Column 2 contains the values of x, because that is the function of x in the second term of the equation.

The data matrix for a parabolic fit would have three columns. The first two columns would be the same as for the straight-line fit. The third column, however, would contain the square of each x value. If, on the other hand, we chose an equation like:

$$\ln p = A + \frac{B}{t} + C \ln t$$

then the first column of the data matrix would contain the value of 1, the second column would contain the reciprocal of the data, and the third column would contain the logarithm of the data. The data vector in this case would have the logarithm of p.

The rectangular data matrix is converted to a square matrix by using a simple operation. The transpose of the data matrix is multiplied by the matrix itself to produce the coefficient matrix. For the straight-line fit of five sets of x-y data, the operation is:

$$\begin{bmatrix} 1 & 1 & 1 & 1 & 1 \\ x_1 & x_2 & x_3 & x_4 & x_5 \end{bmatrix} \begin{bmatrix} 1 & x_1 \\ 1 & x_2 \\ 1 & x_3 \\ 1 & x_4 \\ 1 & x_5 \end{bmatrix}$$

The result is a 2-by-2 matrix containing the required sums of x:

$$\begin{bmatrix} n & \Sigma x \\ \Sigma x & \Sigma x^2 \end{bmatrix}$$

The product of the data vector (considered as a row vector) and the data matrix:

$$\begin{bmatrix} y_1 & y_2 & y_3 & y_4 & y_5 \end{bmatrix} \begin{bmatrix} 1 & x_1 \\ 1 & x_2 \\ 1 & x_3 \\ 1 & x_4 \\ 1 & x_5 \end{bmatrix}$$

gives the required constant vector of length 2:

$$\begin{bmatrix} \Sigma y & \Sigma xy \end{bmatrix}$$

We will now examine a Pascal implementation of this approach. We will also want to incorporate into the program a general technique for calculating the standard error. As we predicted in Chapter 4, the Gauss-Jordan method of solving simultaneous equations will prove to be an important tool to use here because it supplies the *inverse* of the coefficient matrix.

PASCAL PROGRAM: THE MATRIX APPROACH TO CURVE FITTING

Figure 7.3 gives a curve-fitting program that utilizes this matrix approach. The data matrix and data vector are set up at the beginning of procedure LINFIT. Then, procedure SQUARE converts the data matrix,

X, and data vector, $\mathbf{y}$, into the square coefficient matrix, A, and the constant vector, $\mathbf{g}$. The operations are:

$$X^T X = A$$

and

$$\mathbf{y}X = \mathbf{g}$$

The data vector is multiplied by the data matrix to produce the constant vector. The solution vector:

$$A^{-1}\mathbf{g} = \mathbf{b}$$

can be obtained by using any of the routines we developed in Chapter 4 for the solution of simultaneous equations.

The linear curve fitting program developed in Chapter 5 presented the standard errors along with the corresponding elements of the solution to the approximating function. The standard errors were obtained from the standard error of the estimate and the summation of x and x^2. We will use a more general technique at this point.

The standard errors are readily obtained from the inverse of the coefficient matrix, A. The value corresponding to the ith term of the approximating function is the product of the standard error of the estimate (SEE) and the square root of the ith term of the major diagonal of the inverse:

$$\begin{bmatrix} a_{11} & & \\ & a_{22} & \\ & & a_{33} \end{bmatrix}^{-1}$$

The Pascal expression is

$$\text{SIG[I]} := \text{SEE} * \text{SQRT(A[I,I])}$$

Of the several methods we considered previously for the simultaneous solution of linear equations, only the Gauss-Jordan method generated the inverse of the coefficient matrix. Since we need this inverse to determine the errors on the elements of the solution vector, the Gauss-Jordan method is the natural choice. Consequently, we will use this method for the remaining curve-fitting programs in this chapter.

The standard errors on the coefficients can be used to determine the confidence intervals for the corresponding coefficients. In addition, the standard errors can alert us to the possibility of ill conditioning. We discussed ill-conditioned matrices in Chapter 4. They are more likely to be encountered during the solution of simultaneous equations than in curve fitting. Nevertheless, we will want to watch for such a problem.

The standard errors are derived from the square roots of the diagonal elements of the inverted matrix. Large differences in these elements suggest ill conditioning. Therefore, if the squares of the errors are many orders apart, then ill conditioning may be present. The last program in this chapter demonstrates ill conditioning.

The program shown in Figure 7.3 is similar to the one given in Figure 7.1, but the solution to the parabolic equation is determined by the more general matrix method. Three previously developed routines are utilized. The matrix multiplication procedure SQUARE (given in Chapter 3) is shown in the listing. In addition, the Gauss-Jordan procedure GAUSSJ (developed in Chapter 4) and procedure PLOT are referenced as separate files.

```
PROGRAM LEAST2(OUTPUT);
(* Dec 26.1, 1980 *)
(* Pascal program to perform a *)
(* linear least—squares fit *)
(* with Gauss—Jordan routine *)
(* Separate procedures needed:
            GAUSSJ, PLOT *)

CONST
    MAXR = 20; (* data points *)
    MAXC = 4; (* polynomial terms *)

TYPE
    ARY    = ARRAY[1..MAXR] OF REAL;
    ARYS   = ARRAY[1..MAXC] OF REAL;
    ARY2   = ARRAY[1..MAXR, 1..MAXC] OF REAL;
    ARY2S  = ARRAY[1..MAXC, 1..MAXC] OF REAL;

VAR
    X, Y, Y_CALC,
    RESID          : ARY;
    COEF, SIG      : ARYS;
    NROW, NCOL    : INTEGER;
    CORREL_COEF : REAL;
```

Figure 7.3: A Parabolic Least-Squares Fit Using Gauss-Jordan Elimination

```
PROCEDURE GET _ DATA
        (VAR X      : ARY; (* independent variable *)
         VAR Y      : ARY; (* dependent variable *)
         VAR NROW : INTEGER); (* length of vectors *)

VAR
   I: INTEGER;
BEGIN
   NROW := 9;
   FOR I := 1 TO NROW DO
      X[I] := I;
   Y[1] := 2.07; Y[2] := 8.6;
   Y[3] := 14.42; Y[4] := 15.80;
   Y[5] := 18.92; Y[6] := 17.96;
   Y[7] := 12.98; Y[8] := 6.45;
   Y[9] := 0.27
END (* procedure get _ data *);

PROCEDURE WRITE _ DATA;
(* print out the answers *)

VAR
   I: INTEGER;

BEGIN
   WRITELN;
   WRITELN;
   WRITELN(' I   X    Y    Y CALC   RESID');
   FOR I := 1 TO NROW DO
      WRITELN(I: 3, X[I]: 8: 1, Y[I]: 9: 2,
         Y _ CALC[I]: 9: 2, RESID[I]: 9: 2);
   WRITELN;
   WRITELN('coefficients   errors');
   WRITELN(COEF[1], ' ', SIG[1], ' Constant term');
```

Figure 7.3: A Parabolic Least-Squares Fit Using Gauss-Jordan Elimination (cont.)

```
    FOR I := 2 TO NCOL DO
        WRITELN(COEF[I], '   ', SIG[I]) (* other terms *);
    WRITELN;
    WRITELN
        (' Correlation coefficient is ', CORREL_COEF: 8: 5)
END (* write_data *);
PROCEDURE SQUARE(X : ARY2;
                 Y : ARY;
             VAR A : ARY2S;
             VAR G : ARYS;
             NROW,NCOL : INTEGER);
(* matrix multiplication routine *)
(* A = transpose X times X *)
(* G = Y times X *)
VAR
    I, K, L: INTEGER;
BEGIN (* square *)
    FOR K := 1 TO NCOL DO
        BEGIN
            FOR L := 1 TO K DO
                BEGIN
                    A[K,L] := 0.0;
                    FOR I := 1 TO NROW DO
                        BEGIN
                            A[K,L] := A[K,L] + X[I,L] * X[I,K];
                            IF K <> L THEN A[L,K] := A[K,L]
                        END
                END (* L loop *);
            G[K] := 0.0;
            FOR I := 1 TO NROW DO
                G[K] := G[K] + Y[I] * X[I,K]
        END (* k loop *)
END (* square *);
```

Figure 7.3: A Parabolic Least-Squares Fit Using Gauss-Jordan Elimination (cont.)

```
(* PROCEDURE gaussj
    (VAR b     : ary2s;
         y     : arys;
     VAR coef  : arys;
         ncol  : integer;
     VAR error : boolean);
extern; *)
(*$F GAUSSJ.PAS *)


PROCEDURE LINFIT(X,        (* independent variable *)
                  Y : ARY; (* dependent variable *)
     VAR Y_CALC: ARY; (* calculated dep. variable *)
     VAR RESID    : ARY; (* array of residuals *)
     VAR  COEF    : ARYS; (* coefficients *)
     VAR  SIG     : ARYS; (* errors on coefficients *)
          NROW    : INTEGER; (* length of ary *)
     VAR  NCOL    : INTEGER); (* number of terms *)


(* least—squares fit to *)
(* nrow sets of x and y pairs of points *)
(* Separate procedures needed:
    SQUARE — form square coefficient matrix
    GAUSSJ — Gauss—Jordan elimination *)

VAR
    XMATR : ARY2; (* data matrix *)
    A     : ARY2S; (* coefficient matrix *)
    G     : ARYS; (* constant vector *)
    ERROR : BOOLEAN;
    I, J, NM : INTEGER;
    XI, YI, YC, SRS, SEE,
    SUM_Y, SUM_Y2: REAL;
```

Figure 7.3: A Parabolic Least-Squares Fit Using Gauss-Jordan Elimination (cont.)

```
BEGIN (* procedure linfit *)
   NCOL := 3; (* number of terms *)
   FOR I := 1 TO NROW DO
      BEGIN (* setup X matrix *)
         XI := X[I];
         XMATR[I, 1] := 1.0 (* first column *);
         XMATR[I, 2] := XI (* second column *);
         XMATR[I, 3] := XI * XI (* third column *)
      END;
   SQUARE(XMATR, Y, A, G, NROW, NCOL);
   GAUSSJ(A, G, COEF, NCOL, ERROR);
   SUM_Y := 0.0;
   SUM_Y2 := 0.0;
   SRS := 0.0;
   FOR I := 1 TO NROW DO
      BEGIN
         YI := Y[I];
         YC := 0.0;
         FOR J := 1 TO NCOL DO
            YC := YC + COEF[J] * XMATR[I, J];
         Y_CALC[I] := YC;
         RESID[I] := YC - YI;
         SRS := SRS + SQR(RESID[I]);
         SUM_Y := SUM_Y + YI;
         SUM_Y2 := SUM_Y2 + YI * YI
      END;
   CORREL_COEF :=
      SQRT(1.0 - SRS/(SUM_Y2 - SQR(SUM_Y)/NROW));
   IF NROW = NCOL THEN NM := 1
   ELSE NM := NROW - NCOL;
   SEE := SQRT(SRS/NM);
   FOR I := 1 TO NCOL DO (* errors on solution *)
      SIG[I] := SEE * SQRT(A[I, I])
END (* linfit *);
```

Figure 7.3: A Parabolic Least-Squares Fit Using Gauss-Jordan Elimination (cont.)

```
(* PROCEDURE plot(x, y, z: ary; nrow: integer);
extern; *)
(*$F PLOT.PAS *)


BEGIN (* main program *)
   GET_DATA(X, Y, NROW);
   LINFIT(X, Y, Y_CALC, RESID, COEF, SIG, NROW, NCOL);
   WRITE_DATA;
   PLOT(X, Y, Y_CALC, NROW)
END.
```

Figure 7.3: A Parabolic Least-Squares Fit Using Gauss-Jordan Elimination (cont.)

Running the Program

Alter the first program in this chapter or create a new one so that it looks like the one given in the listing of Figure 7.3. Be sure to have procedures GAUSSJ and PLOT available as external files. Alternatively, incorporate these procedures directly into the source program. Run the program and compare the output with Figure 7.4. The results should be the same as for the previous program except that the residuals and errors are given in the new version.

```
     I        X        Y      Y CALC     RESID
     1       1.0      2.07     1.68      -0.39
     2       2.0      8.60     9.02       0.42
     3       3.0     14.42    14.20      -0.22
     4       4.0     15.80    17.21       1.41
     5       5.0     18.92    18.05      -0.87
     6       6.0     17.96    16.73      -1.23
     7       7.0     12.98    13.24       0.26
     8       8.0      6.45     7.58       1.13
     9       9.0      0.27    -0.24      -0.51

   coefficients      errors
   -7.82657e0      1.29802e0   Constant term
    1.05900e1      5.96000e-1
   -1.08295e0      5.81267e-2

   Correlation coefficient is  0.99155
```

Figure 7.4: Output: An Alternate Version of a Parabolic Least-Squares Curve Fit

In the next section the matrix approach will permit us to try out different orders of polynomial equations to fit any given set of data. To develop a sense of the full power of this tool, we will run this new program several times on one set of data. By comparing the resulting plotted curves and correlation coefficients, we will find the best polynomial order to fit our data.

PASCAL PROGRAM:
ADJUSTING THE ORDER OF THE POLYNOMIAL

One of the advantages of the new version of our curve-fitting program is that it is easy to change both the number of rows, corresponding to the number of data points, and the number of columns, corresponding to the number of polynomial terms in the curve-fitting equation. Consequently, for this third version, we will proceed one step further in "generalizing" our program. We will input the order of the polynomial equation from the console. The order, of course, is one smaller than the number of terms in the equation. Make a copy of the previous program and alter it so that it looks like the program shown in Figure 7.5. Changes need to be made to the main program and to procedure LINFIT.

```
PROGRAM LEAST3(INPUT, OUTPUT);
(* Dec 26, 1980 *)
(* Pascal program to perform a *)
(* linear least—squares fit *)
(* with Gauss—Jordan routine *)
(* Separate procedures needed:
          GAUSSJ, PLOT *)
CONST
    MAXR = 20; (* data points *)
    MAXC = 4; (* polynomial terms *)
TYPE
    ARY    = ARRAY[1..MAXR] OF REAL;
    ARYS   = ARRAY[1..MAXC] OF REAL;
    ARY2   = ARRAY[1..MAXR, 1..MAXC] OF REAL;
    ARY2S  = ARRAY[1..MAXC, 1..MAXC] OF REAL;
```

Figure 7.5: Inputting the Polynomial Order from the Console

```
VAR
  X, Y, Y_CALC,
  RESID         : ARY;
  COEF, SIG     : ARYS;
  NROW, NCOL   : INTEGER;
  CORREL_COEF : REAL;
  DONE         : BOOLEAN;


PROCEDURE GET_DATA
            (VAR X      : ARY; (* independent variable *)
             VAR Y      : ARY; (* dependent variable *)
             VAR NROW : INTEGER); (* length of vectors *)

VAR
  I: INTEGER;

BEGIN
  NROW := 9;
  FOR I := 1 TO NROW DO
    X[I] := I;
  Y[1] := 2.07; Y[2] := 8.6;
  Y[3] := 14.42; Y[4] := 15.80;
  Y[5] := 18.92; Y[6] := 17.96;
  Y[7] := 12.98; Y[8] := 6.45;
  Y[9] := 0.27
END (* procedure get_data *);


PROCEDURE WRITE_DATA;
(* print out the answers *)

VAR
  I: INTEGER;
```

Figure 7.5: Inputting the Polynomial Order from the Console (cont.)

```
BEGIN
   WRITELN;
   WRITELN;
   WRITELN(' I    X     Y    Y CALC   RESID');
   FOR I := 1 TO NROW DO
      WRITELN(I: 3, X[I]: 8: 1, Y[I]: 9: 2,
         Y_CALC[I]: 9: 2, RESID[I]: 9: 2);
   WRITELN;
   WRITELN('coefficients   errors');
   WRITELN(COEF[1], ' ', SIG[1], ' Constant term');
   FOR I := 2 TO NCOL DO
      WRITELN
      (COEF[I], ' ', SIG[I]) (* other terms *);
   WRITELN;
   WRITELN
      (' Correlation coefficient is ', CORREL_COEF: 8: 5)
END (* write_data *);

(* PROCEDURE square(x : ary2;
                    y : ary;
                VAR a : ary2s;
                VAR g : arys;
                  nrow,ncol : integer);
extern; *)
(*$F SQUARE.PAS *)

PROCEDURE gaussj
   (VAR b    : ary2s;
        y    : arys;
    VAR coef : arys;
        ncol : integer;
    VAR error : boolean);
extern; *)
(*$F GAUSSJ.PAS *)
```

Figure 7.5: Inputting the Polynomial Order from the Console (cont.)

```
PROCEDURE LINFIT(X,        (* independent variable *)
                   Y : ARY; (* dependent variable *)
           VAR Y_CALC: ARY; (* calculated dep. variable *)
           VAR RESID    : ARY; (* array of residuals *)
           VAR  COEF    : ARYS; (* coefficients *)
           VAR  SIG     : ARYS; (* errors on coefficients *)
                   NROW  : INTEGER; (* length of ary *)
           VAR  NCOL   : INTEGER); (* number of terms *)
(* least—squares fit to *)
(* nrow sets of x and y pairs of points *)
(* Separate procedures needed:
    SQUARE — form square coefficient matrix
    GAUSSJ — Gauss—Jordan elimination *)
VAR
    XMATR  : ARY2; (* data matrix *)
    A          : ARY2S; (* coefficient matrix *)
    G          : ARYS; (* constant vector *)
    ERROR   : BOOLEAN;
    I, J, NM : INTEGER;
    XI, YI, YC, SRS, SEE,
    SUM_Y, SUM_Y2: REAL;
BEGIN (* procedure linfit *)
    FOR I := 1 TO NROW DO
       BEGIN (* setup X matrix *)
          XI := X[I];
          XMATR[I, 1] := 1.0; (* first column *)
          FOR J := 2 TO NCOL DO (* other columns *)
             XMATR[I, J] := XMATR[I,J—1] * XI
       END;
    SQUARE(XMATR, Y, A, G, NROW, NCOL);
    GAUSSJ(A, G, COEF, NCOL, ERROR);
    SUM_Y := 0.0;
    SUM_Y2 := 0.0;
    SRS := 0.0;
```

Figure 7.5: Inputting the Polynomial Order from the Console (cont.)

```
   FOR I := 1 TO NROW DO
      BEGIN
         YI := Y[I];
         YC := 0.0;
         FOR J := 1 TO NCOL DO
            YC := YC + COEF[J] * XMATR[I, J];
         Y_CALC[I] := YC;
         RESID[I] := YC - YI;
         SRS := SRS + SQR(RESID[I]);
         SUM_Y := SUM_Y + YI;
         SUM_Y2 := SUM_Y2 + YI * YI
      END;
   CORREL_COEF :=
      SQRT(1.0 - SRS/(SUM_Y2 - SQR(SUM_Y)/NROW));
   IF NROW = NCOL THEN NM := 1
   ELSE NM := NROW - NCOL;
   SEE := SQRT(SRS/NM);
   FOR I := 1 TO NCOL DO (* errors on solution *)
      SIG[I] := SEE * SQRT(A[I, I])
END (* linfit *);
(* PROCEDURE plot(x, y, z: ary; nrow: integer);
extern; *)
(*$F PLOT.PAS *)
BEGIN (* main program *)
   DONE := FALSE;
   WRITELN;
   GET_DATA (X, Y, NROW);
   REPEAT
      REPEAT
         WRITE( ' Order of polynomial fit? ');
         READLN( NCOL)
      UNTIL NCOL < 5;
      IF NCOL < 1 THEN DONE := TRUE (* quit if ncol < 1 *)
      ELSE
```

Figure 7.5: Inputting the Polynomial Order from the Console (cont.)

```
        BEGIN
            NCOL := NCOL + 1 (* order is one less *);
            LINFIT(X, Y, Y_CALC, RESID, COEF, SIG, NROW, NCOL);
            WRITE_DATA;
            PLOT (X, Y, Y_CALC, NROW)
        END (* ELSE *)
    UNTIL DONE
END.
```

Figure 7.5: Inputting the Polynomial Order from the Console (cont.)

Comparing Runs of the Program

Run the program and input a value of 2 for the polynomial order. The result should again be the same as for the two previous versions. This time, however, the program will cycle and ask for the polynomial order again. Give a value of 1 the second time. The results (shown in Figure 7.6) are the best straight line through the curved set of data.

The two coefficients represent the equation:

$$y = 12.028 - 0.240x$$

Notice that about one-half of the data points are on one side of the straight line and one-half are on the other. The straight line goes through the points, as best it can. The correlation coefficient, however, is only 0.096. This relatively small value indicates that the straight-line fit is not very good.

Give an order of three for the third cycle. This will produce a fit corresponding to the cubic equation:

$$y = -7.2 + 10.0x - 0.946x^2 - 0.00915x^3$$

From the value of the correlation coefficient, it can be seen that the resulting curve describes the data no better than the parabolic fit. One should always use the lowest-order equation that reasonably fits the data. Therefore, the parabola would be the ideal choice in this case.

With this present version, we can choose a polynomial with order up to 3. This restriction is necessary because we have set the maximum number of matrix columns, the constant MAXC, to 4. If a higher order polynomial is necessary, change MAXC to a correspondingly higher value.

The program is terminated by entering a polynomial order that is

```
    I       X        Y       Y CALC      RESID
    1      1.0      2.07      11.79        9.72
    2      2.0      8.60      11.55        2.95
    3      3.0     14.42      11.31       -3.11
    4      4.0     15.80      11.07       -4.73
    5      5.0     18.92      10.83       -8.09
    6      6.0     17.96      10.59       -7.37
    7      7.0     12.98      10.35       -2.63
    8      8.0      6.45      10.11        3.66
    9      9.0      0.27       9.87        9.60

coefficients      errors
 1.20275e1        5.26361     Constant term
-2.39500e-1       9.35368e-1

 Correlation coefficient is   0.09633

    1.00     +                        *
      -
    2.00                  +           *
      -
    3.00                        *        +
      -
    4.00                        *            +
      -
    5.00                        *                 +
      -
    6.00                        *             +
      -
    7.00                        *      +
      -
    8.00              +         *
      -
    9.00  +                     *
          ^         ^         ^         ^         ^         ^
         0.3       4.0       7.7      11.5      15.2      18.9
```

Figure 7.6: A Straight-Line Fit to Parabolic Data

zero or negative. You may want to alter the program so that it asks for the number of terms, rather than the polynomial order. In that case, the statement:

NCOL := NCOL + 1; (* order is one less *)

in the main program should be removed.

In the next three sections we will look at actual experimental applications of curve fitting involving the heat capacity of oxygen, the vapor pressure of liquid lead, and the properties of superheated steam.

PASCAL PROGRAM: THE HEAT-CAPACITY EQUATION

Heat capacity is a measure of how much the temperature of a body will increase when a given amount of heat is added. Experimentally determined data are commonly fitted to the equation:

$$C_p = A + BT + \frac{C}{T^2}$$

where C_p is the heat capacity in units of energy per degree, and T is the absolute temperature. The coefficients are A, B, and C as usual.

Make a copy of the curve-fitting program given in Figure 7.3 (rather than Figure 7.5, which is designed for adjusting the order). Alter procedure GET_DATA and procedure LINFIT so they look like the version shown in Figure 7.7. As before, procedure LINFIT fills the first column of the data matrix with the value of unity, and the second column with the temperature (the independent variable). But now the third column contains the reciprocal of the temperature squared.

The data represent the heat capacity of oxygen over the temperature range of 300 to 1200 kelvins. Notice that X and Y in procedure GET_DATA have been changed to T and CP. These names are local to procedure GET_DATA, and so they can be changed to whatever is convenient.

```
PROCEDURE GET_DATA
               (VAR T      : ARY; (* independent variable *)
                VAR CP     : ARY; (* dependent variable *)
                VAR NROW : INTEGER); (* length of vectors *)
VAR
   I: INTEGER;
BEGIN
   NROW := 10;
   FOR I := 1 TO NROW DO
     T[I] := (I + 2) * 100;
   CP[1] := 7.02; CP[2] := 7.20;
   CP[3] := 7.43; CP[4] := 7.67;
   CP[5] := 7.88; CP[6] := 8.06;
   CP[7] := 8.21; CP[8] := 8.34;
   CP[9] := 8.44; CP[10] := 8.53
END (* procedure get_data *);
```

Figure 7.7: Procedures GET_DATA and LINFIT for the Heat-Capacity Equation

```
PROCEDURE LINFIT(X,         (* independent variable *)
                 Y : ARY; (* dependent variable *)
        VAR Y_CALC: ARY; (* calculated dep. variable *)
        VAR RESID    : ARY; (* array of residuals *)
        VAR  COEF    : ARYS; (* coefficients *)
        VAR  SIG     : ARYS; (* errors on coefficients *)
             NROW    : INTEGER; (* length of ary *)
        VAR  NCOL    : INTEGER); (* number of terms *)
(* least—squares fit to *)
(* nrow sets of x and y pairs of points *)
(* Separate procedures needed:
    SQUARE — form square coefficient matrix
    GAUSSJ — Gauss—Jordan elimination *)

VAR
    XMATR : ARY2; (* data matrix *)
    A        : ARY2S; (* coefficient matrix *)
    G        : ARYS; (* constant vector *)
    ERROR  : BOOLEAN;
    I, J, NM : INTEGER;
    XI, YI, YC, SRS, SEE,
    SUM_Y, SUM_Y2: REAL;

BEGIN (* procedure linfit *)
    NCOL := 3 (* number of terms *);
    FOR I := 1 TO NROW DO
       BEGIN (* setup X matrix *)
          XI := X[I];
          XMATR[I, 1] := 1.0 (* first column *);
          XMATR[I, 2] := XI (* second column *);
          XMATR[I, 3] := 1.0/SQR(XI) (* third column *)
       END;
```

Figure 7.7: Procedures GET_DATA and LINFIT for the Heat-Capacity Equation (cont.)

```
      SQUARE(XMATR, Y, A, G, NROW, NCOL);
      GAUSSJ(A, G, COEF, NCOL, ERROR);
      SUM _ Y := 0.0;
      SUM _ Y2 := 0.0;
      SRS := 0.0;
      FOR I := 1 TO NROW DO
        BEGIN
          YI := Y[I];
          YC := 0.0;
          FOR J := 1 TO NCOL DO
            YC := YC + COEF[J] * XMATR[I, J];
          Y _ CALC[I] := YC;
          RESID[I] := YC − YI;
          SRS := SRS + SQR(RESID[I]);
          SUM _ Y := SUM _ Y + YI;
          SUM _ Y2 := SUM _ Y2 + YI * YI
        END;
      CORREL _ COEF := =
            SQRT(1.0 − SRS/(SUM _ Y2 − SQR(SUM _ Y)/NROW));
      IF NROW = NCOL THEN NM := 1
      ELSE NM := NROW − NCOL;
      SEE := SQRT(SRS/NM);
      FOR I := 1 TO NCOL DO (* errors on solution *)
        SIG[I] := SEE * SQRT(A[I, I])
    END (* linfit *);
```

Figure 7.7: Procedures GET _ DATA and LINFIT for the Heat-Capacity Equation (cont.)

Compile the program and run it. The output should look like Figure 7.8. The resulting equation is:

$$C_p = 6.9 + 0.00143T - \frac{32610}{T^2}$$

PASCAL PROGRAM: THE VAPOR PRESSURE EQUATION

When a gas or vapor is in equilibrium with its own liquid or solid, we say that it is saturated. For pure materials, the saturation pressure is a single-valued function of the temperature. A commonly used equation to express the relationship between the saturation pressure and the saturation temperature is:

$$\log P = A + \frac{B}{T} + C \log T$$

In this equation, P is the pressure and T is the absolute temperature. The coefficients are A, B, and C. Either the natural logarithm or the common logarithm is used.

```
    I      X         Y      Y CALC     RESID
    1    300.0      7.02     6.97      -0.05
    2    400.0      7.20     7.27       0.07
    3    500.0      7.43     7.49       0.06
    4    600.0      7.67     7.67       0.00
    5    700.0      7.88     7.84      -0.04
    6    800.0      8.06     8.00      -0.06
    7    900.0      8.21     8.16      -0.05
    8   1000.0      8.34     8.31      -0.03
    9   1100.0      8.44     8.46       0.02
   10   1200.0      8.53     8.60       0.07

 coefficients       errors
 6.90485e0         1.29216e-1   Constant term
 1.43448e-3        1.26803e-4
-3.26106e4         1.16750e4

 Correlation coefficient is  0.99482

  300.00 *+
    -
  400.00          +   *
    -
  500.00               +*
    -
  600.00                    *
    -
  700.00                        *+
    -
  800.00                           *  +
    -
  900.00                              *+
    -
 1000.00                                  *+
    -
 1100.00                                    +*
    -
 1200.00                                       +   *
         ^        ^          ^         ^        ^        ^
        7.0      7.3        7.6       8.0      8.3      8.6
```

Figure 7.8. Output: The Heat Capacity of Oxygen

Make a copy of the program shown in Figure 7.7. Alter procedures GET_DATA and LINFIT so that they look like the versions shown in Figure 7.9. Procedure LINFIT sets the first column of the matrix to 1 as usual. Column 2 is then filled with the reciprocal of the independent variable, the temperature. Column 3 gets the logarithm of the temperature.

```
PROCEDURE GET_DATA
            (VAR T      : ARY; (* independent variable *)
             VAR P      : ARY; (* dependent variable *)
             VAR NROW : INTEGER); (* length of vectors *)

VAR
   I: INTEGER;

BEGIN (* get_data *)
   NROW := 10;
   FOR I := 1 TO NROW DO
      T[I] := (I + 6.0) * 100.0;
   P[1] := 1.0E−9; P[2] := 5.598E−8;
   P[3] := 1.234E−6; P[4] := 1.507E−5;
   P[5] := 1.138E−4; P[6] := 6.067E−4;
   P[7] := 2.512E−3; P[8] := 8.337E−3;
   P[9] := 2.371E−2; P[10] := 5.875E−2;
   FOR I := 1 TO NROW DO
      P[I] := LN(P[I]) (* take log of p *)
END (* procedure get_data *);

PROCEDURE LINFIT(X,          (* independent variable *)
                 Y : ARY; (* dependent variable *)
      VAR Y_CALC: ARY; (* calculated dep. variable *)
      VAR RESID    : ARY; (* array of residuals *)
      VAR  COEF    : ARYS; (* coefficients *)
      VAR  SIG     : ARYS; (* errors on coefficients *)
           NROW  : INTEGER; (* length of ary *)
      VAR  NCOL   : INTEGER); (* number of terms *)
```

Figure 7.9: Procedures GET_DATA and LINFIT for the Vapor Pressure Equation

```
(* least—squares fit to *)
(* nrow sets of x and y pairs of points *)
(* Separate procedures needed:
    SQUARE — form square coefficient matrix
    GAUSSJ — Gauss—Jordan elimination *)

VAR
    XMATR  : ARY2; (* data matrix *)
    A      : ARY2S; (* coefficient matrix *)
    G      : ARYS; (* constant vector *)
    ERROR  : BOOLEAN;
    I, J, NM: INTEGER;
    XI, YI, YC, SRS, SEE,
    SUM_Y, SUM_Y2: REAL;
BEGIN (* procedure linfit *)
    NCOL := 3 (* number of terms *);
    FOR I := 1 TO NROW DO
        BEGIN (* setup X matrix *)
            XI := X[I];
            XMATR[I, 1] := 1.0 (* first column *);
            XMATR[I, 2] := 1.0/XI (* second column *);
            XMATR[I, 3] := LN(XI) (* third column *)
        END;
    SQUARE(XMATR, Y, A, G, NROW, NCOL);
    GAUSSJ(A, G, COEF, NCOL, ERROR);
    SUM_Y := 0.0;
    SUM_Y2 := 0.0;
    SRS := 0.0;
    FOR I := 1 TO NROW DO
        BEGIN
            YI := Y[I];
            YC := 0.0;
```

Figure 7.9: Procedures GET _ DATA and LINFIT for the Vapor Pressure Equation (cont.)

```
        FOR J := 1 TO NCOL DO
           YC := YC + COEF[J] * XMATR[I, J];
        Y_CALC[I] := YC;
        RESID[I] := YC - YI;
        SRS := SRS + SQR(RESID[I]);
        SUM_Y := SUM_Y + YI;
        SUM_Y2 := SUM_Y2 + YI * YI
     END;
  CORREL_COEF :=
     SQRT(1.0 - SRS/(SUM_Y2 - SQR(SUM_Y)/NROW));
  IF NROW = NCOL THEN NM := 1
  ELSE NM := NROW - NCOL;
  SEE := SQRT(SRS/NM);
  FOR I := 1 TO NCOL DO (* errors on solution *)
     SIG[I] := SEE * SQRT(A[I, I])
END (* linfit *);
```

Figure 7.9: Procedures GET _ DATA and LINFIT for the Vapor Pressure Equation (cont.)

Run the new program. The results should look like Figure 7.10.

```
     I      X          Y        Y CALC    RESID
     1    700.0     -20.72     -20.72    -0.00
     2    800.0     -16.70     -16.70    -0.01
     3    900.0     -13.61     -13.59     0.02
     4   1000.0     -11.10     -11.11    -0.01
     5   1100.0      -9.08      -9.09    -0.00
     6   1200.0      -7.41      -7.41     0.00
     7   1300.0      -5.99      -5.99    -0.01
     8   1400.0      -4.79      -4.78     0.00
     9   1500.0      -3.74      -3.74     0.00
    10   1600.0      -2.83      -2.83     0.00

    coefficients      errors
     1.81432e1       5.96613e-1   Constant term
    -2.31700e4       7.77388e1
    -8.80333e-1      7.48010e-2

    Correlation coefficient is  1.00000
```

Figure 7.10: Output: The Vapor Pressure of Lead

The data represent the vapor pressure of liquid lead over the temperature range of 700 to 1600 kelvins. The corresponding equation is:

$$\ln P = 18.1 - \frac{23170}{T} - 0.880 \ln T$$

when the pressure is in atmospheres and the natural logarithm is used. The correlation coefficient of 1.00000 indicates a very good fit.

Notice that the original pressures in procedure GET_DATA are given in atmospheres. These values are then converted to the logarithm of the pressure in procedure GET_DATA. The x and y values that are printed are actually the temperature and the logarithm of the pressure. It is left as an exercise for the reader to preserve the original pressure values. An additional array, which might be called PRESS, can be used for this purpose. Then procedure WRITE_DATA can be altered so that it prints both the original pressure and the logarithm of the pressure.

In the final example we will encounter an equation that breaks the restriction we established at the beginning of the chapter. One of the unknown coefficients is in a nonlinear form—specifically, an exponential. Later, in Chapter 10, we will study algorithms for handling such an equation, but here we will have to be satisfied with a less elegant solution. We will estimate values for the exponent until we find the optimum correlation coefficient.

A THREE-VARIABLE EQUATION

In the previous section, we considered an equation of state for a saturated gas. The pressure is a function of temperature for this condition. We will now consider an equation of state for a superheated gas. In this case, the temperature is above the saturation temperature or, put another way, the pressure is below the saturation pressure. Since temperature and pressure are independent variables under this condition, we can express the volume as a function of both the temperature and the pressure.

For an ideal gas, the equation of state is:

$$PV = RT$$

where V is the molar volume, T is the temperature, P is the pressure, and R is the gas constant. But as the pressure increases, the behavior of the gas becomes less and less ideal. A common non-ideal equation of state is:

$$PV = A + BP + CP^2 + DP^3 + \ldots$$

It can be seen that this equation is just a power series expansion in pressure.

Be careful not to confuse the gas constant (R) with the residuals (r). In the Pascal program, the symbol R is used for the gas constant and the symbol RESID is used for the residuals.

Since all gases become ideal as the pressure is reduced, the equation of state should merge smoothly with the ideal gas equation as the pressure approaches zero. For this reason, the value of A in the above equation must be equal to RT. Because fewer polynomial terms are needed at lower values of pressure, the equation of state can often be written as:

$$PV = RT + BP + CP^2$$

The determination of the coefficients B and C is straightforward when the temperature is constant. However, if temperature is also a variable, then the coefficients B and C need to be functions of temperature.

Several different equations of state are in common use. All of them, however, are empirical. As an example of a three-variable equation, consider the expression:

$$PV = RT + \frac{BP}{T^n} + CP^2$$

In this case, coefficient C is not a function of T, but the original coefficient B has become the function:

$$\frac{B}{T^n}$$

This equation has a nonlinear coefficient, n, and so we cannot obtain a solution by the methods discussed in the chapter. However, if an estimate is made for the coefficient, *n*, then the remaining linear coefficients can be determined. We will begin with an estimate of unity for coefficient *n* and determine the other coefficients. We can then observe how well the resulting equation represents the original data. Then we will change the value of *n* and see whether the new equation is better or worse.

PASCAL PROGRAM: AN EQUATION OF STATE FOR STEAM

The program given in Figure 7.11 can be used to find the coefficients *B* and *C* for the published properties of steam. Since the coefficient of the first term on the right is unity, the equation has been rearranged to give:

$$PV - RT = \frac{BP}{T^n} + CP^2$$

The data are defined in procedure GET _ DATA. The temperature is given in degrees Fahrenheit, the pressure in pounds per square inch, and the specific volume in cubic feet per pound mass. The temperature data are converted to the absolute Rankine scale by the addition of 460. Pressures are left in pounds per square inch. But then the gas constant, *R*, which has a value of 85.76 for steam, is multiplied by 144 square inches per square foot.

The data matrix is set up in procedure LINFIT, as usual. The first column of the matrix contains the pressure divided by the *n*th power of the temperature. Column 2 contains the square of the pressure. The **y** vector has the value *PV − RT*.

Notice that this is the first time that we did *not* put the value of unity in the first column of the matrix. We could, of course, divide the equation by the pressure. The right-hand side would then look like a first-order, straight-line fit. But this might cause trouble when pressure became very small.

Comparing Runs of the Program to Determine the Value of *n*

Type up the program given in Figure 7.11 and execute it. Notice that the exponent *n* is initially chosen to be unity. The results should look like Figure 7.12.

```
PROGRAM LEAST6(OUTPUT);
(* Feb 8, 81 *)
(* Pascal program to perform a *)
(* linear least—squares fit *)
(* on the properties of steam *)
(* with Gauss—Jordan routine *)
(* Separate procedures needed:
          GAUSSJ, PLOT *)
CONST
    MAXR = 20 (* data points *);
    MAXC = 4 (* polynomial terms *);
TYPE
    ARY   = ARRAY[1..MAXR] OF REAL;
    ARYS  = ARRAY[1..MAXC] OF REAL;
    ARY2  = ARRAY[1..MAXR, 1..MAXC] OF REAL;
    ARY2S = ARRAY[1..MAXC, 1..MAXC] OF REAL;
VAR
    P, T, V, Y,
    Y_CALC, RESID : ARY;
    COEF, SIG     : ARYS;
    NROW, NCOL    : INTEGER;
    CORREL_COEF  : REAL;
PROCEDURE GET_DATA
            (VAR P, T   : ARY; (* independent variables *)
             VAR V      : ARY; (* dependent variable *)
             VAR NROW  : INTEGER); (* length of vectors *)
VAR
    I: INTEGER;
BEGIN
    NROW := 12;
    T[1] := 400; P[1] := 120; V[1] := 4.079;
    T[2] := 450; P[2] := 120; V[2] := 4.36;
    T[3] := 500; P[3] := 120; V[3] := 4.633;
```

Figure 7.11: An Equation of State for Steam

```
        T[4] := 400; P[4] := 140; V[4] := 3.466;
        T[5] := 450; P[5] := 140; V[5] := 3.713;
        T[6] := 500; P[6] := 140; V[6] := 3.952;
        T[7] := 400; P[7] := 160; V[7] := 3.007;
        T[8] := 450; P[8] := 160; V[8] := 3.228;
        T[9] := 500; P[9] := 160; V[9] := 3.44;
        T[10] := 400; P[10] := 180; V[10] := 2.648;
        T[11] := 450; P[11] := 180; V[11] := 2.85;
        T[12] := 500; P[12] := 180; V[12] := 3.042;
        FOR I := 1 TO NROW DO
            T[I] := T[I] + 460.0 (* convert to Rankine *)
END (* procedure get_data *);


PROCEDURE WRITE_DATA;
(* print out the answers *)
VAR
    I: INTEGER;
BEGIN
    WRITELN;
    WRITELN(' I   P   T   V  ',
                    'Y   Y CALC   %RES');
    FOR I := 1 TO NROW DO
        WRITELN(I:3, P[I]:7:1, T[I]:7:1, V[I]:7:3,
        Y[I]:9:2, Y_CALC[I]:9:2, (100.0 * RESID[I]/Y[I]):9:2);
    WRITELN;
    WRITELN('coefficients   errors');
    WRITELN(COEF[1], ' ', SIG[1], ' Constant term');
    FOR I := 2 TO NCOL DO
        WRITELN(COEF[I], ' ', SIG[I]) (* other terms *);
    WRITELN;
    WRITELN
        (' Correlation coefficient is ', CORREL_COEF: 8: 5)
END (* write_data *);
```

Figure 7.11: An Equation of State for Steam (cont.)

```
(* PROCEDURE square(x : ary2;
                    y : ary;
               VAR a : ary2s;
               VAR g : arys;
              nrow,ncol : integer);
extern; *)
(*$F SQUARE.PAS *)

(* PROCEDURE gaussj
   (VAR b    : ary2s;
        y    : arys;
    VAR coef : arys;
        ncol : integer;
    VAR error : boolean);
extern; *)
(*$F GAUSSJ.PAS *)


PROCEDURE LINFIT
           (P, T, V   :         (* independent variables *)
       VAR Y         : ARY; (* dependent variable *)
       VAR Y_CALC: ARY; (* calculated dep. variable *)
       VAR RESID    : ARY; (* array of residuals *)
       VAR  COEF    : ARYS; (* coefficients *)
       VAR  SIG     : ARYS; (* errors on coefficients *)
            NROW  : INTEGER; (* length of ary *)
       VAR  NCOL   : INTEGER); (* number of terms *)
(* fit an equation of state through
   nrow sets of p, t, and v sets of points *)
(* Separate procedures needed:
   SQUARE — form square coefficient matrix
   GAUSSJ — Gauss—Jordan elimination *)
CONST
   R = 85.76 (* gas constant for steam *);
```

Figure 7.11: An Equation of State for Steam (cont.)

```
VAR
    XMATR : ARY2; (* data matrix *)
    A       : ARY2S; (* coefficient matrix *)
    G       : ARYS; (* constant vector *)
    ERROR : BOOLEAN;
    I, J, NM : INTEGER;
    POWER, YI, YC, SRS,
    SEE, SUM_Y, SUM_Y2: REAL;
BEGIN (* procedure linfit *)
    NCOL := 2 (* number of terms *);
    FOR I := 1 TO NROW DO
        BEGIN (* setup X matrix *)
            POWER := T[I];
            XMATR[I, 1] := P[I]/POWER (* first column *);
            XMATR[I, 2] := SQRT(P[I]) (* second column *);
            Y[I] := V[I] * P[I] — R * T[I]/144.0
        END;
    SQUARE(XMATR, Y, A, G, NROW, NCOL);
    GAUSSJ(A, G, COEF, NCOL, ERROR);
    SUM_Y := 0.0;
    SUM_Y2 := 0.0;
    SRS := 0.0;
    FOR I := 1 TO NROW DO
        BEGIN
            YI := Y[I];
            YC := 0.0;
            FOR J := 1 TO NCOL DO
                YC := YC + COEF[J] * XMATR[I, J];
            Y_CALC[I] := YC;
            RESID[I] := YC — YI;
            SRS := SRS + SQR(RESID[I]);
            SUM_Y := SUM_Y + YI;
            SUM_Y2 := SUM_Y2 + YI * YI
        END;
```

Figure 7.11: An Equation of State for Steam (cont.)

```
    CORREL_COEF :=
        SQRT(1.0 - SRS/(SUM_Y2 - SQR(SUM_Y)/NROW));
    IF NROW = NCOL THEN NM := 1
    ELSE NM := NROW - NCOL;
    SEE := SQRT(SRS/NM);
    FOR I := 1 TO NCOL DO (* errors on solution *)
        SIG[I] := SEE * SQRT(A[I, I])
END (* linfit *);
BEGIN (* main program *)
    GET_DATA(P, T, V, NROW);
    LINFIT(P, T, V, Y, Y_CALC, RESID, COEF, SIG, NROW, NCOL);
    WRITE_DATA
END.
```

Figure 7.11: An Equation of State for Steam (cont.)

```
    I     P       T       V       Y       Y CALC    %RES
    1   120.0   860.0   4.079   -22.70   -19.44   -14.36
    2   120.0   910.0   4.360   -18.76   -17.43    -7.07
    3   120.0   960.0   4.633   -15.77   -15.63    -0.92
    4   140.0   860.0   3.466   -26.94   -24.16   -10.30
    5   140.0   910.0   3.713   -22.14   -21.82    -1.44
    6   140.0   960.0   3.952   -18.45   -19.72     6.85
    7   160.0   860.0   3.007   -31.06   -28.98    -6.69
    8   160.0   910.0   3.228   -25.48   -26.30     3.24
    9   160.0   960.0   3.440   -21.33   -23.90    12.03
   10   180.0   860.0   2.648   -35.54   -33.88    -4.67
   11   180.0   910.0   2.850   -28.96   -30.86     6.59
   12   180.0   960.0   3.042   -24.17   -28.16    16.50

    coefficients       errors
    -2.62187e2        4.75869e1   Constant term
     1.56514e0        6.49499e-1

    Correlation coefficient is 0.91839
```

Figure 7.12: Output: The Properties of Superheated Steam (the exponent n = 1)

It can be seen from Figure 7.12 that the results are not very good. The correlation coefficient is 92%, but some of the calculated values are more than 10% from the original data. We should therefore try something else. Alter the variable POWER near the beginning of procedure LINFIT so that the exponent n will be 2. That is, make POWER equal to the square of the temperature.

POWER : = SQR(T[I]);

Rerun the program. The output should look like Figure 7.13.

The resulting curve fit looks better this time. The correlation coefficient is 98%, and the calculated points are within 6% of the data. We will now continue in this way by increasing the exponent to 3. Change POWER to read:

POWER : = T[I] ∗ SQR(T[I]);

Run the program a third time and compare the output to Figure 7.14.

I	P	T	V	Y	Y CALC	%RES
1	120.0	860.0	4.079	-22.70	-21.49	-5.32
2	120.0	910.0	4.360	-18.76	-18.03	-3.85
3	120.0	960.0	4.633	-15.77	-15.10	-4.25
4	140.0	860.0	3.466	-26.94	-26.01	-3.44
5	140.0	910.0	3.713	-22.14	-21.98	-0.71
6	140.0	960.0	3.952	-18.45	-18.56	0.58
7	160.0	860.0	3.007	-31.06	-30.59	-1.49
8	160.0	910.0	3.228	-25.48	-25.98	2.00
9	160.0	960.0	3.440	-21.33	-22.08	3.48
10	180.0	860.0	2.648	-35.54	-35.22	-0.88
11	180.0	910.0	2.850	-28.96	-30.04	3.74
12	180.0	960.0	3.042	-24.17	-25.64	6.08

```
coefficients      errors
-1.99359e5        1.19367e4   Constant term
 9.90932e-1       1.80298e-1

Correlation coefficient is  0.98901
```

Figure 7.13: Output: The Properties of Superheated Steam (the exponent n = 2)

```
    I     P      T      V       Y      Y CALC    %RES
    1   120.0  860.0  4.079   -22.70   -22.88    0.78
    2   120.0  910.0  4.360   -18.76   -18.80    0.22
    3   120.0  960.0  4.633   -15.77   -15.52   -1.59
    4   140.0  860.0  3.466   -26.94   -26.97    0.13
    5   140.0  910.0  3.713   -22.14   -22.21    0.35
    6   140.0  960.0  3.952   -18.45   -18.39   -0.32
    7   160.0  860.0  3.007   -31.06   -31.09    0.10
    8   160.0  910.0  3.228   -25.48   -25.65    0.68
    9   160.0  960.0  3.440   -21.33   -21.28   -0.23
   10   180.0  860.0  2.648   -35.54   -35.22   -0.90
   11   180.0  910.0  2.850   -28.96   -29.10    0.49
   12   180.0  960.0  3.042   -24.17   -24.19    0.06

 coefficients      errors
 -1.38664e8       1.50165e6  Constant term
  2.99897e-1      2.52008e-2

 Correlation coefficient is 0.99963
```

Figure 7.14: Output: The Properties of Superheated Steam (the exponent n = 3)

The result is definitely better. The fitted values are within one percent of the original points and the correlation coefficient is 99.96 percent. It looks as though we should accept these results. But just to be sure, change the exponent to 4 with the statement:

POWER := SQR(SQR(T[I]));

Run the program again and compare the results to Figure 7.15. It can be seen that we have definitely gone too far. The calculated points are not as close as they were for the previous fit and the correlation coefficient is farther from unity. In fact, the Gauss-Jordan procedure may report that the matrix is singular. The problem is ill conditioning. This can be seen by comparing the squares of the two standard errors. This ratio is greater than twenty orders of magnitude.

At this point, we could try to improve the fit by choosing noninteger exponents close to the value of 3. The results would show, however, that 3 is the best value. Another possibility is to make coefficient C a function of temperature. But since the result with an exponent of 3 is reasonably good, we should perhaps be satisfied.

```
     I    P      T       V        Y      Y CALC    %RES
     1  120.0  860.0   4.079    -22.70   -23.69    4.38
     2  120.0  910.0   4.360    -18.76   -19.32    3.02
     3  120.0  960.0   4.633    -15.77   -16.00    1.45
     4  140.0  860.0   3.466    -26.94   -27.46    1.94
     5  140.0  910.0   3.713    -22.14   -22.36    1.02
     6  140.0  960.0   3.952    -18.45   -18.49    0.19
     7  160.0  860.0   3.007    -31.06   -31.22    0.51
     8  160.0  910.0   3.228    -25.48   -25.39   -0.34
     9  160.0  960.0   3.440    -21.33   -20.96   -1.74
    10  180.0  860.0   2.648    -35.54   -34.96   -1.62
    11  180.0  910.0   2.850    -28.96   -28.41   -1.89
    12  180.0  960.0   3.042    -24.17   -23.43   -3.08

  coefficients       errors
  -9.84751e10        3.64805e9  Constant term
  -1.90703e-1        6.82835e-2

  Correlation coefficient is 0.99571
```

Figure 7.15: Output: The Properties of Superheated Steam (the exponent n = 4)

SUMMARY

This chapter has illustrated the programming concept of *generalization*. We have now progressed through several versions of our curve-fitting program:

- a straight-line fit
- a parabolic fit
- a more direct parabolic fit, using the matrix approach
- a general polynomial curve fit in which the polynomial order can be adjusted
- curve fits for nonpolynomial equations.

The restriction on all of these versions is that the coefficients must be linear. Although we found a way around this restriction in the last example of the chapter, we must still develop methods of dealing with nonlinear coefficients.

CHAPTER 8

Solution of Equations by Newton's Method

INTRODUCTION

In this chapter, we will develop a computer program for solving equations by a technique known as Newton's method, or the Newton-Raphson method. This method is especially suitable for finding particular roots of "well-behaved" functions. We will use Newton's method in Chapter 10 when we return to the topic of curve fitting with nonlinear coefficients. Of course, as we will see, Newton's method has many other applications and would be worth our attention even if we did not plan to use it as a tool later in this book.

First we will outline the mathematical formulation of Newton's method; then we will consider a series of progressively more significant Pascal implementations of the technique. We will study two pitfalls of the method—the case where the tangent to the curve has a zero slope, and the case where successive approximations fail to converge on a root. We will then consider ways of dealing with these pitfalls in our program. Also, we will encounter the problematic Pascal SIN function that we discussed in Chapter 1. Finally, we will use our program to solve a practical application—the vapor pressure equation.

FORMULATING NEWTON'S METHOD

Let us begin by considering a general equation of the form:

$$f(x) = 0 \tag{1}$$

This equation might have one solution, several solutions, or none at all. That is, there may be particular values of x that make the equation equal to zero. These values are called the *roots* or *solutions* of the equation. For other values of x, the function will not be zero.

Sometimes we can solve such an equation explicitly. For example, the expression:

$$x^2 - 4 = 0$$

can be converted to:

$$x^2 = 4$$

which has the solutions:

$$x = 2 \quad \text{and} \quad x = -2$$

But sometimes, an equation cannot be solved so easily. As an example, consider the expression:

$$\ln P = A + \frac{B}{T} + C \ln T$$

This formula, which we will implement into our program at the end of this chapter, can be used to describe the vapor pressure of a material. In this equation P is the pressure, T is the temperature, and A, B, and C are constants that are unique for each substance. For the element lead, the experimentally determined coefficients are

$$A = 18.19$$
$$B = -23180$$
$$C = -0.8858$$

when the pressure is given in atmospheres and the temperature is given in degrees Kelvin.

We can easily find the vapor pressure of lead at, say, 1000° by solving the expression:

$$\ln P = A + \frac{B}{1000} + C \ln 1000$$

But suppose that we want to find the temperature that corresponds to a lead vapor pressure of 0.1 atmospheres. We want to solve the equation:

$$\ln 0.1 = 18.19 - \frac{23180}{T} - 0.8858 \ln T$$

This nonlinear equation cannot be solved explicitly for the temperature. However, we can use an approximation method to calculate the answer to as high a precision as we desire.

For the general case, we write the equation:

$$y = f(x)$$

We are thus interested in the values of x when y is equal to zero; that is, we want to determine the points where the curve of the function crosses the x-axis.

Consider, for example, the curve of the equation:

$$y = f(x) = x^2 - 4$$

This curve crosses the x-axis at two places, $+2$ and -2, as shown in Figure 8.1.

As a second example, consider the curve of:

$$y = f(x) = x^2$$

This curve is tangent to the x-axis at the origin, corresponding to a single root, $x = 0$, as shown in Figure 8.2.

Finally, consider the equation:

$$y = f(x) = x^2 + 4$$

shown in Figure 8.3. This equation does not cross the x-axis at all, and so it has no real roots.

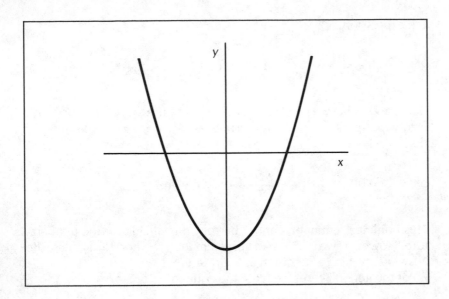

Figure 8.1: A Function with Two Solutions

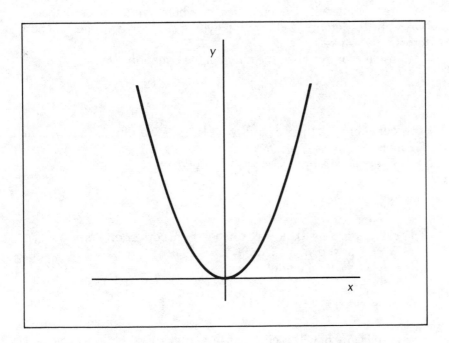

Figure 8.2: A Function with One Solution

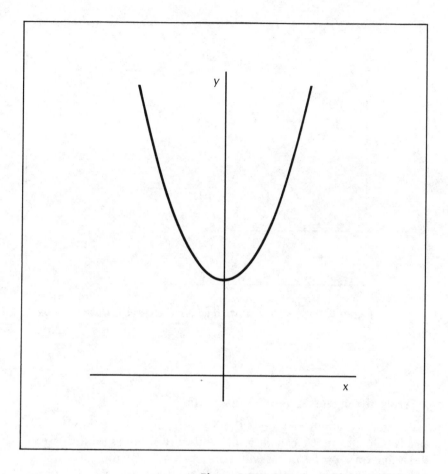

Figure 8.3: A Function with No Real Roots

Let us explore the behavior of a general function,

$$y = f(x)$$

in the vicinity of a root. We might find that it looks like the curve of Figure 8.4.

The function crosses the x-axis at a root because the relationship:

$$y = f(x) = 0$$

is satisfied there.

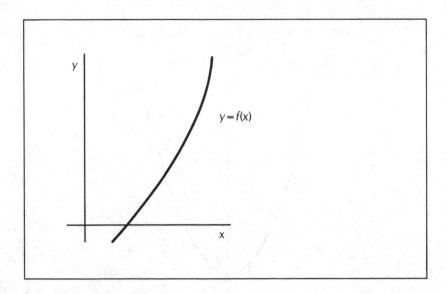

Figure 8.4: f(x) = 0 is satisfied where the curve crosses the x-axis.

A Series of Tangents to the Curve $y = f(x)$

We start Newton's method with an approximate value for x, say x_1, that is near a root. We can determine the corresponding value of y by the equation $y_1 = f(x_1)$. This will represent a point on the curve that is not, in general, a root. A tangent to $f(x)$ is now constructed at this point on the curve. The tangent is extended until it intersects the x-axis. The next approximation, x_2, is at this intersection on the x-axis, as illustrated in Figure 8.5. Notice that in this example, the second approximation, x_2, is closer to the root than the first approximation, x_1. Thus we have refined our original approximation.

The process is now repeated. The function is evaluated at $x = x_2$ to obtain the corresponding value of y, $y_2 = f(x_2)$. The value of y_2 is smaller than the value of y_1, indicating that we are closer to the root. A tangent is again constructed, this time at the point $(x_2, f(x_2))$. The intersection of the new tangent with the x-axis gives the value of x_3, the third approximation of x. We continue in this way, improving the value of x until we are as close to the actual root as we want.

Let us go back and review the first step in more detail. The initial approximation, x_1, gives rise to $y_1 = f(x_1)$. The tangent constructed at y_1

has a slope of:

$$f'(x_1) = \frac{y_1}{x_1 - x_2} \tag{2}$$

Because $y_1 = f(x_1)$, Equation 2 can be expressed as:

$$x_2 = x_1 - \frac{f(x_1)}{f'(x_1)} \tag{3}$$

or more generally as:

$$x_{i+1} = x_i - \frac{f(x_i)}{f'(x_i)} \tag{4}$$

where x_i is the ith approximation. Equation 4 is the usual form of Newton's method. It can be an ideal technique for finding a desired root for a real-life equation.

There are potential problems with the use of Newton's method, which we will consider later in this chapter. But the equations that deal with the behavior of real things typically have only one meaningful

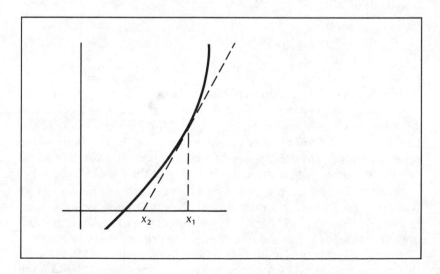

Figure 8.5: The tangent crosses the x-axis closer to the root than the original approximation for x.

root. The other roots of such equations will usually be negative, zero, or complex. Furthermore, the approximate value of the answer may be known. For example, the ideal-gas law can provide a first approximation to a more complicated equation of state.

Now that we have arrived at an equation for the general form of Newton's method, developing our program will be relatively easy.

PASCAL PROGRAM: A FIRST ATTEMPT AT NEWTON'S METHOD

We will implement Newton's method for a simple problem, one for which we already know the answer. The equation we will solve is:

$$x^2 = 2$$

or

$$x^2 - 2 = 0 \tag{5}$$

The positive solution to this equation is, of course, the square root of 2. First, we define the function:

$$y = f(x) = x^2 - 2 \tag{6}$$

and its derivative:

$$\frac{dy}{dx} = f'(x) = 2x \tag{7}$$

Our first attempt at a program for Newton's method is shown in Figure 8.6. The algorithm itself is contained in procedure NEWTON. A separate procedure, called FUNC, is used for the calculation of the function (Equation 6) and its derivative (Equation 7). The body of the main program provides the first approximation. It calls procedure NEWTON to find the solution and to print out the answer.

As a matter of good programming practice, procedures and functions that are used to calculate values should not print intermediate results. But in this particular case, it is instructive to observe the successive values of x, $f(x)$ and $f'(x)$ during the convergence procedure. Consequently, there is a print statement in procedure NEWTON for this purpose. You may want to remove this print statement when the program is working properly.

```
PROGRAM NEWDR(OUTPUT);
(* Version 1, Jan 23, 81 *)
VAR
   X, X2      : REAL;
   ALLDONE    : BOOLEAN;
   ERROR      : BOOLEAN;

PROCEDURE FUNC(X : REAL;
        VAR FX, DFX : REAL);

BEGIN
   FX  := X * X − 2.0;
   DFX := 2.0 * X
END (* func *);

PROCEDURE NEWTON
           (VAR X: REAL);
CONST
   TOL = 1.0E−6;

VAR
   FX, DFX, DX, X1 : REAL;

BEGIN (* newton *)
   REPEAT
      X1 := X;
      FUNC(X, FX, DFX);
      DX := FX/DFX;
      X := X1 − DX;
      WRITELN
      ('x= ', X1, ', fx= ', FX, ', dfx= ', DFX)
   UNTIL  ABS(DX) <= ABS(TOL * X)
END (* newton *);
```

Figure 8.6: Newton's Method, Version One

```
BEGIN (* main program *)
   WRITELN;
   X := 2.0 (* first guess *);
   NEWTON(X);
   WRITELN;
   WRITELN('The solution is ', X);
   WRITELN
END.
```

Figure 8.6: Newton's Method, Version One (cont.)

Tolerance

There are a few additional matters we should consider at this time. Successive approximations are provided by a **REPEAT/UNTIL** loop in procedure NEWTON. This loop continues until two successive values are within the desired tolerance. We are not interested in whether the difference (DX) between two successive approximations has a negative or a positive value. We are only concerned with the magnitude. For this reason, we must be careful to take the absolute value of the comparison.

Furthermore, we are not interested in the actual difference, but only the relative difference. Suppose, for example, that we want our answers to be accurate to one part in a million, that is, 1 part in 10^6. If a particular solution has a value of unity, then two successive values must be closer than 10^{-6}. If, however, the solution itself has a value of 10^{-6}, then two successive values must differ by no more than 10^{-12}. We therefore choose a relative criterion rather than an absolute criterion for termination of the iteration process.

Generalizing Procedure Calls

Another matter we should consider is the relationship of procedure NEWTON to the procedure it calls. This is procedure FUNC that is used to calculate the function and its derivative. Procedure NEWTON gives directions for carrying out the operation described by Equation 4. It is independent of the actual function it is operating on. For this reason, procedure FUNC, used to obtain the function and its derivative, should be an entirely separate entity.

Ideally, the procedure name FUNC, within NEWTON, should be a dummy parameter. The actual procedure name should be passed to procedure NEWTON as an actual parameter during execution. With this arrangement, procedure NEWTON could be directed to solve one equation at one point and another equation at another point. This technique is easily accomplished in FORTRAN. However, Pascal compilers, as currently implemented, generally do not allow procedure names to be passed as parameters to other procedures.

If two or more different equations are to be solved in the same program, we will have to provide additional, separate copies of procedure NEWTON. Each copy will, of course, be given a different name. Another, more complicated approach would be to use a **CASE** statement in procedure FUNC. This approach would select one function on the first call, a second function on the next call, and so on.

Running the Program

Type up the program shown in Figure 8.6 and try it out. The first approximation for the square root of 2 is chosen to be 2. When the program is executed, it should produce the square root of 2 after several iterations. The results will look something like Figure 8.7.

```
        x= 2.00000, fx= 2.00000,   dfx= 4.00000
        x= 1.50000, fx= 2.50000E-1, dfx= 3.00000
        x= 1.41667, fx= 6.94442E-3, dfx= 2.83333
        x= 1.41422, fx= 5.96046E-6, dfx= 2.82843
        x= 1.41421, fx=-1.19209E-7, dfx= 2.82843

    The solution is  1.41421
```

Figure 8.7: Output: The Positive Root of f(x) = x² − 2

In the following sections we will make several small changes to refine this program. The first change will allow us to input different first-approximation values for the root. This facility is important for studying equations that have more than one root.

Adding User Input for the First Approximation

When the first version of Newton's method is working properly, we can begin to add new features. For the remaining versions in this chapter there will be both console input and output. Consequently, the first line of the source program should be changed to:

PROGRAM NEWDR2 (INPUT, OUTPUT);

Make a copy of the first version. Alter the main program, (the part within the final **BEGIN/END** block) so that it looks like Figure 8.8.

```
BEGIN (* main program *)
   ALLDONE := FALSE;
   REPEAT
      WRITELN;
      WRITE(' First guess: ');
      READLN( X);
      IF X < -19.0 THEN ALLDONE := TRUE
      ELSE
         BEGIN
            NEWTON(X);
            WRITELN;
            WRITELN('The solution is ', X);
            WRITELN
         END
   UNTIL ALLDONE
END.
```

Figure 8.8: The Main Program for Version Two

Compile the new version and try it out. For the first version, the value of 2 was used as the first guess to Newton's method. The second version is more sophisticated than the first. The user is asked to input the first guess. The successive approximations, along with the function and its derivative, are displayed as before. At the conclusion of the task, the program begins again and the user is asked to input another first approximation.

Running the Program to Find the Second Root

Start with the value of 2; the results should be the same as for the first version. Then, for the second cycle, try the value of 1. This first approximation is on the other side of the root, but the square root of 2 should again be found in relatively few steps. Try the value of − 2 for the third cycle. Notice that the process converges on a different root this time.

There are, of course, two solutions to the equation:

$$x^2 - 2 = 0$$

We found the other root this time by giving a negative first approximation.

Investigate what happens when the first guess is near the midpoint of the two roots. Try a first guess of 0.0001. In this case, the process takes quite a few steps to produce the answer. Finally, try a first guess of zero. The curve

$$y = f(x)$$

has zero slope at this point. Consequently, one of two things can happen. Either a floating-point divide error can occur or the program can loop indefinitely. This problem will be corrected in the next version.

A Test for Zero Slope

When the derivative, or slope, of our function is zero, the final term in Equation 4 becomes infinite. We are looking for the point where the slope crosses the x-axis. But it can be seen from Figure 8.9 that these two lines are parallel. Consequently, they do not intersect.

We will now add some instructions for testing the slope to our Newton's method program. One way to do this is to define a small number such as:

CONST SMALL = 1.0E−16;

Then, the slope can be tested with the statement:

IF ABS(DFX) < SMALL **THEN** . . .

One problem with this approach, however, is that the value assigned to SMALL must be consistent with the particular version of Pascal being used. That is, the value of SMALL may have to be chosen carefully.

The check for zero slope that we will actually use is more straightforward:

IF DFX = 0.0 **THEN** . . .

If the slope is found to be zero, an error message is printed and an error flag is set. Otherwise, the process continues normally. The second change to our program occurs in the main driver. Locate the line:

WRITELN('The solution is ', X);

near the end of the program. Slide this line over two spaces and insert the following line immediately before it:

IF NOT ERROR **THEN**

The two lines will look like this:

IF NOT ERROR **THEN**
    WRITELN('The solution is ', X);

The purpose of this new line is to test the error flag after each call to procedure NEWTON. If the flag is not set, then there is no error and the solution is printed. Alter your procedure NEWTON so that is looks like Figure 8.10. Compile the new version and try it out.

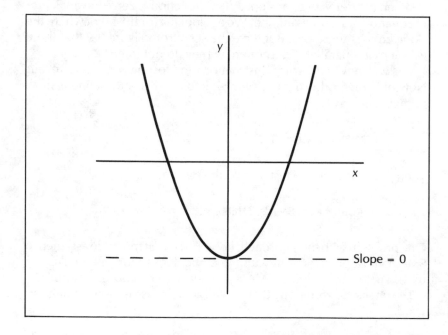

Figure 8.9: At f'(x) = 0 the tangent is parallel to the x-axis.

```
PROCEDURE NEWTON
          (VAR X: REAL);
 (* Jan 23, 81 *)

CONST
   TOL = 1.0E−6;

VAR
   FX, DFX, DX, X1 : REAL;

BEGIN (* newton *)
   ERROR := FALSE;
   REPEAT
     X1 := X;
     FUNC(X, FX, DFX);
     IF DFX = 0.0 THEN
        BEGIN
           ERROR := TRUE;
           X := 1.0;
           WRITELN('ERROR: slope zero')
        END
     ELSE
        BEGIN
           DX := FX/DFX;
           X := X1 − DX;
           WRITELN
               ('x= ', X1, ', fx= ', FX, ', dfx= ', DFX)
        END
   UNTIL
     ERROR OR
     (ABS(DX) <= ABS(TOL * X))
END (* newton *);
```

Figure 8.10: Newton's Method with a Test for Zero Slope

Running the Program with the Slope Test

Give initial values 2, 1, and −1 as before to see that the program behaves properly. Then, give an initial value of zero. This input caused a floating-point divide check in the previous version. Now however, the Pascal program should handle this input with no difficulty. The program should print the appropriate error message, then request another first approximation. The program can be aborted by entering a first approximation that is less than −19.

Finally, our next task is to make sure the program will print an error message and terminate after an appropriate number of iterations if approximations do not converge on a root.

Failure to Converge

Sometimes, Newton's method will not converge on a root after a reasonable number of iterations. One possibility is that successive approximations are oscillating around a complex root, as illustrated in Figure 8.11.

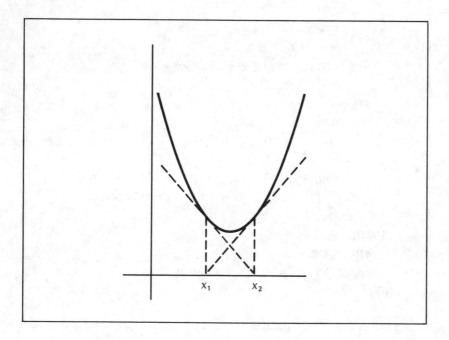

Figure 8.11: A Complex Root

The first approximation, x_1, gives rise to the second value, x_2. But x_2 then produces the original value of x_1 for the third approximation. In this case, the process will never terminate.

Another possibility is that an approximation is very far from a root. This can occur even though the first guess appears to be close to a root. We can observe this behavior with our previous version of Newton's method if we run the program again and give a first approximation of 0.00001. This will produce a second value of 20,000, which is quite an overshoot. Each successive approximation will now be about one-half of the previous value. A solution will eventually be obtained in this case; but it will take more than 20 iterations before it converges.

We can guard against lack of convergence by adding a loop counter to procedure NEWTON. We can then abort the procedure if convergence does not occur after, say, 20 iterations. This will take care of the case of oscillation, as well as the case of an approximation that is very far from a root.

Alter procedure NEWTON so that it looks like Figure 8.12. Compile the new version and try it out. Give an initial approximation of 500000. This first guess is so far from the root that it will take many cycles to converge. The iteration process should stop after 20 loops and the appropriate error message will be displayed. Next, try a first guess of 2 to make sure that everything is still all right.

We have developed and refined our program using a simple, predictable function. Now we are ready to use our program to find the roots of some more difficult functions.

PASCAL PROGRAMS: SOLVING OTHER EQUATIONS

Consider the nonlinear equation:

$$e^x = 4x$$

This equation, unlike our previous one, cannot be solved explicitly for x. Consequently, it is a suitable candidate for our Newton's method program. The corresponding function, and its derivative are:

$$f(x) = e^x - 4x$$

$$f'(x) = e^x - 4$$

```
PROCEDURE NEWTON
          (VAR X: REAL);
(* Jan 23, 81 *)

CONST
   TOL = 1.0E−6;
   MAX = 20;

VAR
   FX, DFX, DX, X1 : REAL;
   I : INTEGER;

BEGIN (* newton *)
   ERROR := FALSE;
   I := 0;
   REPEAT
      I := I + 1;
      X1 := X;
      FUNC(X, FX, DFX);
      IF DFX = 0.0 THEN
         BEGIN
            ERROR := TRUE;
            X := 1.0;
            WRITELN('ERROR: slope zero')
         END
      ELSE
         BEGIN
            DX := FX/DFX;
            X := X1 − DX;
            WRITELN
               ('x= ', X1, ', fx= ', FX, ', dfx= ', DFX)
         END
```

Figure 8.12: Newton's Method with a Loop Counter

```
    UNTIL
        ERROR OR
        (I > MAX) OR
        (ABS(DX) <= ABS(TOL * X));
    IF I > MAX THEN
        BEGIN
            WRITELN
                ('ERROR: no convergence in ',
                MAX, ' loops');
            ERROR := TRUE
        END
END (* newton *);
```

Figure 8.12: Newton's Method with a Loop Counter (cont.)

Change procedure FUNC so that it looks like Figure 8.13. Notice that the variable E is introduced into procedure FUNC so that the exponent of X will not have to be calculated twice. Recall that

$$\frac{de^x}{dx} = e^x$$

```
PROCEDURE FUNC(X : REAL;
        VAR FX, DFX : REAL);
VAR
    E : REAL;
BEGIN
    E   := EXP(X);
    FX  := E - 4.0 * X;
    DFX := E - 4.0
END (* func *);
```

Figure 8.13: Procedure FUNC for $e^x = 4x$

Run the new version and input a first guess of 4. The program should converge on a value of 2.153. This is one of the two roots. The other root can be found by giving a first guess of 0.1. The result, in this case, will be a value of 0.357.

Because our next equation involves the SIN function, its implementation will require one of two versions, depending on the Pascal compiler in use.

A Function with Many Roots

Let us explore the several roots of the equation:

$$\sin(x) = \frac{x}{10}$$

Change procedure FUNC so that it looks like Figure 8.14.

Zero is one of the roots of the new equation. As we approach this root, we will need to take the SIN of numbers that are closer and closer to zero. Unfortunately, as we discovered in Chapter 1, several of the microprocessor implementations of Pascal contain an error that will cause trouble in this case. The problem is that the argument to SIN is squared before a range check is performed.

If you have not run the test programs given in Chapter 1, you should do so at this time. If the results of these tests indicate that the dynamic range of the SIN function is the same as for other floating-point operations, then you can probably use the procedure FUNC given in Figure 8.14. However, if you find that the dynamic range of the SIN function is approximately the square root of the usual floating-point operations, then you should use the procedure FUNC given in Figure 8.15 instead.

```
PROCEDURE FUNC(X : REAL;
         VAR FX, DFX : REAL);

BEGIN
    FX  := SIN( X ) − 0.1 * X;
    DFX := COS( X ) − 0.1
END (* func *);
```

Figure 8.14: The Solution of sin (x) = x / 10

Tests with several different Pascals produced the following dynamic ranges:

	Sin	General
Pascal 1	10^{-38}	10^{-38}
Pascal 2	10^{-18}	10^{-18}
Pascal 3	10^{-18}	10^{-38}
Pascal 4	10^{-21}	10^{-64}
Pascal 5	10^{-9}	10^{-18}

The alternate approach, given in Figure 8.15, is to inspect the argument to sine and cosine. If the argument is closer to zero than the value of SMALL, then the built-in functions SIN and COS are not called. Instead, procedure FUNC will substitute the argument for the value of sine and unity for the value of cosine in this case. The value of SMALL should be adjusted to fit your version of Pascal. If in doubt, use the version of Figure 8.15.

```
PROCEDURE FUNC(X : REAL;
          VAR FX, DFX : REAL);
CONST
    SMALL = 1.0E−8;
VAR
    S, C : REAL;
BEGIN
    IF ABS(X) > SMALL THEN
        BEGIN
            S := SIN( X );
            C := COS( X )
        END
    ELSE
        BEGIN
            S := X;
            C := 1.0
        END;
    FX  := S − 0.1 * X;
    DFX := C − 0.1
END (* func *);
```

Figure 8.15: Alternate Procedure FUNC for sin (x) = x / 10

Compile this latest version and try it out. Give a first estimate of 1.0. This should converge on the root at zero. Then, try a first guess of – 1.0. This too should converge on the root at zero. There are several other roots to this equation. The table in Figure 8.16 gives first approximations and the corresponding roots. Try each one to verify that your Newton's method is working properly.

1st approximation	root
– 1	0
1	0
4	2.852..
4.3	7.068..
4.5	0
4.7	– 8.423..
5	– 2.852..
6	7.068..
9	8.423..

Figure 8.16: The Roots of f(x) = sin (x) – x / 10

PASCAL PROGRAM: THE VAPOR PRESSURE EQUATION

We are now ready to solve the vapor pressure equation that was introduced at the beginning of this chapter. We will write our function and its derivative as:

$$f(T) = A + \frac{B}{T} + C \ln T - \ln P$$

$$df(T) = \frac{-B}{T^2} + \frac{C}{T}$$

Remember that A, B, C, and P are constants. Make a copy of the Newton program and alter procedure FUNC so that it looks like Figure 8.17. Notice that the letter T is used in place of X. This change is possible since the parameters to the procedure are dummy variables.

```
PROCEDURE FUNC(T : REAL;
          VAR FT, DFT : REAL);

(* the vapor pressure of lead *)

CONST
    A = 18.19;
    B = -23180.0;
    C = -0.8858;
    LOGP = -4.60517 (* ln(.01) *);

BEGIN
    FT := A + B/T + C * LN(T) - LOGP;
    DFT := -B/(T * T) + C/T
END (* func *);
```

Figure 8.17: Solution of the Vapor Pressure Equation

With this FUNC we will find the temperature that corresponds to a vapor pressure of 0.01 atmospheres. The logarithm of this pressure, -4.60517, is entered directly as a constant.

Compile the new version and try it out. Give a first guess of 500 and the program will converge to a temperature of 1416 in about seven steps.

SUMMARY

Using a familiar function at first, we saw how easy it was to write, and then improve upon a Pascal program implementing Newton's method for finding the roots of an equation. We then used our program to solve some functions that are more complex. These included exponential and trigonometric functions. Finally, we solved the equation of an actual scientific application.

CHAPTER 9

Numerical Integration

INTRODUCTION

In this chapter, we will develop three different methods for carrying out numerical integration, that is, for determining the area beneath a curve between two given values of the independent variable. After writing Pascal programs for each of the methods, we will judge how efficiently each method computes the area. Each of the methods uses progressively smaller "panels" of measurable area to divide up the area beneath the curve. The summation of the areas of these panels then provides an approximation of the total area.

In the trapezoidal rule method, these panels are topped by straight-line secants to the curve, whereas Simpson's method tops each panel with a parabolic curve of its own. Both of these methods can be improved with an abbreviated Taylor series expansion to compute the error—a method referred to as *end correction*.

The third integration method we will study, the Romberg method, is the most complex; it constructs a matrix of interpolations to arrive at the total area. One of the functions we will try on both the Simpson and Romberg methods is related to the normal distribution function, which we will be returning to in Chapter 11. Finally, we will explore a method for integrating a function that approaches infinity at one of its limits.

Let us begin with a brief description of the definite integral and the methods of evaluating it.

THE DEFINITE INTEGRAL

The evaluation of the *definite integral*:

$$\int_a^b f(x)dx = F(b) - F(a)$$

where $F'(x) = dF(x)/dx = f(x)$, can be interpreted as the area under the curve of the function $f(x)$ from the *limit* a to the *limit* b, as illustrated in Figure 9.1.

The actual integration can be straightforward for certain functions, but very difficult for others. For example, the power series:

$$1 + 2x + 3x^2$$

can be integrated term by term and then evaluated between the limits of a and b to give:

$$b + b^2 + b^3 - a - a^2 - a^3$$

More complicated functions can sometimes be integrated by reference to integration tables found in math handbooks. Other techniques, such as *integration by parts* can sometimes be used. Another possibility is to replace the original function with an infinite series or an *asymptotic expansion*. The integration is then carried out on the replacement function.

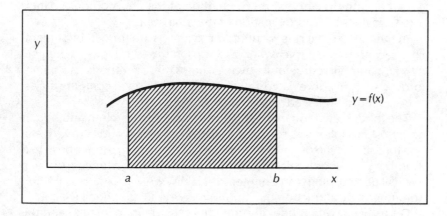

Figure 9.1: The Area under the Curve y = f(x)

Sometimes the integrand is not a proper function at all, but simply a collection of experimental data. In such a case, it may be possible to fit the experimental data to a function that can be integrated. Nevertheless, there will be times when it will be impossible or very difficult to evaluate the integrand analytically. But if the limits are known, then an approximation method may provide an acceptable solution. This approach is known as *numerical integration*.

There are several different methods that are commonly used for numerical integration. These typically involve the substitution of an easily integrated function for the original function. The new function may be a polynomial such as a straight line or parabola, or it may consist of transcendental functions such as sines and cosines. The accuracy of the resulting calculation depends on how well the substituted function approximates the original function.

In the next section we will describe both the general method of function substitution, and a specific, simple implementation of the method—the use of straight-line functions.

THE TRAPEZOIDAL RULE

One of the simplest methods of numerical integration is known as the *trapezoidal rule*. In this method, the original function is approximated by a set of straight lines. The region to be integrated is divided into uniformly spaced sections or panels. The panel width, Δx, is:

$$\Delta x = \frac{b - a}{n}$$

where n is the number of panels and a and b are the integration limits. If the entire integral is fitted with a single straight line, that is, if there is only one panel (as illustrated in Figure 9.2) then the calculated area is:

$$\frac{(b - a)[f(a) + f(b)]}{2}$$

In this formula, $f(a)$ is the value of the function at the left limit and $f(b)$ is the value at the right limit.

For the more general case, the area is divided into n panels. An interior panel is bounded on the left by a vertical line at x_i, and on the right by a vertical line at $x_i + \Delta x$. The lower edge of the panel is marked by the x-axis. The original curve at the top of the panel is replaced by a straight line that in general will not have a zero slope (that is, it will not be parallel to the x-axis). The resulting panel thus has the shape of a trapezoid, giving rise to the name of the method.

The panels can be numbered from 1 to n, but actually we are interested in evaluating the function at the left and right edges of each panel. There are $n - 1$ interior edges for the n panels, in addition to the right and left boundaries of the integral. Consequently, there will be $n + 1$ edges, which can be numbered 0 through n.

The area of the first panel is:

$$\frac{[f(0) + f(1)]\Delta x}{2}$$

or

$$\frac{[f(a) + f(1)]\Delta x}{2}$$

and the area of the last panel is:

$$\frac{[f(n-1) + f(n)]\Delta x}{2}$$

or

$$\frac{[f(n-1) + f(b)]\Delta x}{2}$$

The area of the ith panel is:

$$\frac{[f(i-1) + f(i)]\Delta x}{2}$$

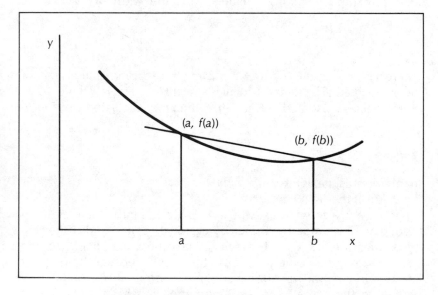

Figure 9.2: Calculation of the Area with One Panel

where $f(i - 1)$ is the value of the function on the left side of the ith panel and $f(i)$ is the value of the function on the right side of the panel. The desired integral is the sum of the areas of all of the panels. Thus the total area can be calculated by summing the areas of all of the panels according to the expression:

$$\{[f(a) + f(1)] + [f(1) + f(2)] + [f(2) + f(3)] + \dots$$
$$+ [f(n - 2) + f(n - 1)] + [f(n - 1) + f(b)]\} \frac{\Delta x}{2}$$

The right edge of the first panel is also the left edge of the second panel, and the left edge of the last panel is also the right edge of the next to last panel. The edges of all of the other panels are also common to two panels. The formula for integration by the trapezoidal rule can therefore be simplified as:

$$[f(a) + 2f(1) + 2f(2) + \dots +$$
$$2f(n - 2) + 2f(n - 1) + f(b)] \frac{\Delta x}{2}$$

Our first computer program for this method will allow us to experiment with the number of panels we use to divide up the total area.

PASCAL PROGRAM:
THE TRAPEZOIDAL RULE WITH USER INPUT FOR THE NUMBER OF PANELS

The area calculated from the trapezoidal method more closely approaches the actual value as the number of panels becomes larger and larger. This can be demonstrated with the program given in Figure 9.3. Type up the program and execute it. The main program will ask the user for the number of sections into which the area is to be divided. The resulting calculated value will then be based on the given number of panels.

The algorithm for integration by the trapezoidal rule is contained in procedure TRAPEZ. The function to be integrated:

$$\int_{1}^{9} \frac{dx}{x}$$

is defined in function FX. This is a separate routine from TRAPEZ. Since this function can be readily integrated, we can compare the exact value

```
PROGRAM TRAP1(INPUT, OUTPUT);

(* integration by the trapezoidal rule *)
(* Dec 24, 80 *)

VAR
    DONE : BOOLEAN;
    SUM, UPPER, LOWER : REAL;
    PIECES : INTEGER;

FUNCTION FX(X : REAL) : REAL;

(* find f(x) = 1/x *)
(* watch out for x = 0 *)

BEGIN
    FX := 1.0/X
END;

PROCEDURE TRAPEZ(LOWER, UPPER : REAL;
                         PIECES  : INTEGER;
                    VAR SUM  : REAL);

(* numerical integration by the trapezoid method *)
(* function is FX, limits are LOWER and UPPER *)
(* with number of regions equal to PIECES *)
(* fixed partition is DELTA_X, answer is SUM *)

VAR
    I : INTEGER;
    X, DELTA_X, ESUM, PSUM : REAL;
```

Figure 9.3: The Trapezoidal Rule

```
BEGIN
    DELTA_X := (UPPER − LOWER)/PIECES;
    ESUM := FX(LOWER) + FX(UPPER);
    PSUM := 0.0;

    FOR I := 1 TO PIECES − 1 DO
    BEGIN
        X := LOWER + I * DELTA_X;
        PSUM := PSUM + FX(X)
    END;
    SUM := (ESUM + 2.0 * PSUM) * DELTA_X * 0.5
END; (* trapez *)

BEGIN (* main program *)
    DONE := FALSE;
    LOWER := 1.0;
    UPPER := 9.0;
    WRITELN;
    REPEAT
        WRITE('How many sections? ');
        READLN(PIECES);
        IF PIECES < 0 THEN DONE := TRUE

        ELSE
            BEGIN
                TRAPEZ(LOWER, UPPER, PIECES, SUM);
                WRITELN(' area = ', SUM)
            END
    UNTIL DONE
END.
```

Figure 9.3: The Trapezoidal Rule (cont.)

of the integral to the value calculated by the trapezoidal method. The value of the integral is the natural logarithm of 9, or 2.197225.

We will now look at a more sophisticated version of this program.

PASCAL PROGRAM: AN IMPROVED TRAPEZOIDAL RULE

We can improve our trapezoidal program in two ways. First, we can change procedure TRAPEZ so that it will automatically divide the original area into more and more pieces. Second, we can avoid much calculation at each step by using the results from the previous step.

Alter the first trapezoidal program so that it looks like the one given in Figure 9.4. Compile the program and execute it.

```pascal
PROGRAM TRAP2(OUTPUT);

(* integration by the trapezoidal rule *)
(* Feb 6, 81 *)

CONST
    TOL = 1.0E−5;

VAR
    SUM, UPPER, LOWER : REAL;

FUNCTION FX(X : REAL) : REAL;

(* find f(x) = 1/x *)
(* watch out for x = 0 *)

BEGIN
    FX := 1.0/X
END;
```

Figure 9.4: An Improved Trapezoidal Method

```
PROCEDURE TRAPEZ(LOWER, UPPER, TOL :  REAL;
                          VAR SUM : REAL);


(* numerical integration by the trapezoid method *)
(* external function is FX *)
(* limits are LOWER and UPPER *)
(* with number of regions equal to PIECES *)
(* partition width is DELTA_X, answer is SUM *)

VAR
    PIECES, I: INTEGER;
    X, DELTA_X, END_SUM, MID_SUM, SUM1: REAL;

BEGIN
    PIECES := 1;
    DELTA_X := (UPPER — LOWER)/PIECES;
    END_SUM := FX(LOWER) + FX(UPPER);
    SUM := END_SUM * DELTA_X/2.0;
    WRITELN('   1', SUM);
    MID_SUM := 0.0;
    REPEAT
       PIECES := PIECES * 2;
       SUM1 := SUM;
       DELTA_X := (UPPER — LOWER)/PIECES;
       FOR I := 1 TO PIECES DIV 2 DO
          BEGIN
             X := LOWER + DELTA_X * (2.0 * I — 1.0);
             MID_SUM := MID_SUM + FX(X)
          END;
       SUM := (END_SUM + 2.0 * MID_SUM) * DELTA_X * 0.5;
       WRITELN(PIECES:5, SUM)
    UNTIL ABS(SUM — SUM1) <= ABS(TOL * SUM)
END; (* trapez *)
```

Figure 9.4: An Improved Trapezoidal Method (cont.)

```
BEGIN (* main program *)
   LOWER := 1.0;
   UPPER := 9.0
   WRITELN;
   TRAPEZ(LOWER, UPPER, TOL, SUM);
   WRITELN;
   WRITELN('   area = ', SUM)
END.
```

Figure 9.4: An Improved Trapezoidal Method (cont.)

Running the Program

For the first calculation, the entire area is taken as a single panel. The number of panels is then doubled to 2 and the new area is calculated. The procedure continues with a doubling of the number of panels at each step. As the number of panels increases, so does the accuracy of the result and, of course, the length of the computation time. The number of panels and the corresponding calculated areas are displayed at each step. The output should look like Figure 9.5.

It can be seen from Figure 9.5 that the calculated value more closely approaches the correct value as the number of panels is increased. The process terminates when two successive values are within the desired tolerance. The tolerance is a relative global variable that is set to the

```
        1  4.44444
        2  3.02222
        4  2.46349
        8  2.27341
       16  2.21733
       32  2.20234
       64  2.19851
      128  2.19755
      256  2.19731
      512  2.19725
     1024  2.19723

     area =   2.19723
```

Figure 9.5: Integration by the Improved Trapezoidal Method (The number of panels and the corresponding areas are given.)

value of 10^{-5} near the top of the program. You may want to change this value to correspond to the precision of your Pascal. You may also want to remove the statement in procedure TRAPEZ that prints the successive values during the iteration process.

With this version, the number of panels at each step is twice the number used for the previous step. Notice that it is not necessary to recalculate all of the panel heights at each step. Suppose, for example, that four panels are used for a particular step. The next step will then use eight panels. Since all of the panel heights from the previous step are common to the present step, it is faster to save the sum from each step and use it for the next step rather than recalculate all the heights. Only the heights midway between the previous points need to be computed. These are then added to the previous sum of the interior values. The new sum is multiplied by 2 and added to the values at each end of the function. This result is then multiplied by one-half the panel width to obtain the area.

While this improved version runs faster than the first version, it suffers from the same errors. We are trying to fit a general curve with a set of straight lines. If the original curve is not a straight line, many steps may be needed for convergence.

In our final version of the trapezoidal rule method we will include a calculation of end correction in our program.

End Correction

We can further improve the trapezoidal method by including correction factors obtained from a Taylor series expansion. The error terms contain second derivatives of the function. Fortunately, however, simplified correction terms can be used instead. The terms for the interior points vanish, leaving only the terms at the upper and lower limits of the function. Thus the major error-correction term requires a derivative of the function at the two limits, x_0 and x_n. The additional quantity to be included in the sum is

$$\frac{[f'(b) - f'(a)](\Delta x)^2}{12}$$

where $f'(b)$ is the value of the slope at b and $f'(a)$ is the value of the slope at a. The resulting value is subtracted from the regular trapezoidal sum. The method is referred to as integration by the trapezoidal rule with end correction.

PASCAL PROGRAM:
TRAPEZOIDAL RULE WITH END CORRECTION

Make a copy of the previous program. Add function DFX (Figure 9.6) immediately after function FX. This procedure will calculate the derivative of the function. We could, of course, combine the calculation of the function and its derivative in a single routine. But we would then need to use a procedure rather than a function. Furthermore, the derivative is only needed at the two end points. Change procedure TRAPEZ according to Figure 9.6.

```
FUNCTION DFX(X : REAL) : REAL;

(* find derivative of f(x) = 1/x *)

BEGIN
    DFX := −1.0/SQR(X)
END;

PROCEDURE TRAPEZ(LOWER, UPPER, TOL : REAL;
                            VAR SUM : REAL);

(* numerical integration by the trapezoid method *)
(* external function is FX *)
(* limits are LOWER and UPPER *)
(* with number of regions equal to PIECES *)
(* partition width is DELTA _ X, answer is SUM *)

VAR
    PIECES,I : INTEGER;
    X, DELTA _ X, END _ SUM, MID _ SUM,
    END _ COR, SUM1: REAL;
```

Figure 9.6: Function DFX and Procedure TRAPEZ with End Correction

```
BEGIN
    PIECES := 1;
    DELTA_X := (UPPER − LOWER)/PIECES;
    END_SUM := FX(LOWER) + FX(UPPER);
    END_COR := (DFX(UPPER) − DFX(LOWER))/12.0;
    SUM := END_SUM * DELTA_X/2.0;
    WRITELN('  1', SUM);
    MID_SUM := 0.0;
    REPEAT
        PIECES := PIECES * 2;
        SUM1 := SUM;
        DELTA_X := (UPPER − LOWER)/PIECES;
        FOR I := 1 TO PIECES DIV 2 DO
            BEGIN
                X := LOWER + DELTA_X * (2.0 * I − 1.0);
                MID_SUM := MID_SUM + FX(X)
            END;
        SUM := (END_SUM + 2.0 * MID_SUM) * DELTA_X
            * 0.5 − SQR(DELTA_X) * END_COR;
        WRITELN(PIECES:5, SUM)
    UNTIL ABS(SUM − SUM1) <= ABS(TOL * SUM)
END (* trapez *);
```

Figure 9.6: Function DFX and Procedure TRAPEZ with End Correction (cont.)

Running the Program

Compile the program and execute it. The results should look like Figure 9.7.

Notice how much faster the process converges in this case. Also notice that during the iteration process, the sum passed the correct value, then reversed its direction. It should be realized, however, that it will not always be possible to use the end correction technique with the trapezoidal method. This will be the case if the slope at one limit or the other is infinite or becomes very large. (We will discuss this case at the end of this chapter.) Also, if the function has a zero slope at the limits, the end-correction term will be zero.

```
         1  4.44444
         2  1.70535
         4  2.13427
         8  2.19111
        16  2.19675
        32  2.19719
        64  2.19722
       128  2.19723

   area  =   2.19723
```

Figure 9.7: Trapezoidal Integration with End Correction

The trapezoidal rule method replaces the function with a first-order polynomial (a straight line). In many cases we can expect a second-order polynomial (a parabola) to be a better substitute; this is the idea of the method we will implement in our next program. We will examine three different runs of this program—first with the same function we ran on the trapezoidal rule program, then with an exponential and a sine function.

PASCAL PROGRAM: SIMPSON'S INTEGRATION METHOD

Simpson's method of numerical integration is similar to the trapezoidal method except that the original curve is replaced by a set of parabolas rather than straight lines. Since a parabola can be defined by a minimum of three points, we must divide the original area into an even number of panels. Each parabola is fitted to the tops of two adjacent panels. In general, we expect parabolas to produce a better fit than straight lines. That is, we should need fewer panels to obtain satisfactory convergence. The formula is:

$$(f_0 + f_n + 4 \sum_{\substack{j=1 \\ j \text{ odd}}}^{n-1} f_j + 2 \sum_{\substack{j=2 \\ j \text{ even}}}^{n-2} f_j) \; \frac{\Delta x}{3}$$

First Run of the Simpson's Rule Program

Type up the program shown in Figure 9.8 and execute it. The results should look like Figure 9.9.

```
PROGRAM simp1(output);

(* integration by Simpson's method *)
(* Dec 24, 80 *)

CONST
   TOL = 1.0E−5;

VAR
   SUM, UPPER, LOWER : REAL;

FUNCTION FX(X : REAL) : REAL;

(* find f(x) = 1/x *)
(* watch out for x = 0 *)

BEGIN
   FX := 1.0/X
END (* function fx *);

PROCEDURE SIMPS(LOWER, UPPER, TOL : REAL;
                          VAR SUM : REAL);

(* numerical integration by Simpson's rule *)
(* function is FX, limits are LOWER and UPPER *)
(* with number of regions equal to PIECES *)
(* partition is DELTA_X, answer is SUM *)

VAR
   I : INTEGER;
   X, DELTA_X, EVEN_SUM,
   ODD_SUM, END_SUM, SUM1 : REAL;
   PIECES : INTEGER;
```

Figure 9.8: Simpson's Rule

```
BEGIN
   PIECES := 2;
   DELTA_X := (UPPER − LOWER)/PIECES;
   ODD_SUM := FX(LOWER + DELTA_X);
   EVEN_SUM := 0.0;
   END_SUM := FX(LOWER) + FX(UPPER);
   SUM := (END_SUM + 4.0 * ODD_SUM) * DELTA_X/3.0;
   WRITELN(PIECES:5, SUM);
   REPEAT
      PIECES := PIECES * 2;
      SUM1 := SUM;
      DELTA_X := (UPPER − LOWER)/PIECES;
      EVEN_SUM := EVEN_SUM + ODD_SUM;
      ODD_SUM := 0.0;
      FOR I := 1 TO PIECES DIV 2 DO
         BEGIN
            X := LOWER + DELTA_X * (2.0 * I − 1.0);
            ODD_SUM := ODD_SUM + FX(X)
         END;
      SUM := (END_SUM + 4.0 * ODD_SUM
            + 2.0 * EVEN_SUM) * DELTA_X/3.0;
      WRITELN(PIECES:5, SUM)
   UNTIL ABS(SUM − SUM1) <= ABS(TOL * SUM)
END (* simps *);

BEGIN (* main program *)
   LOWER := 1.0;
   UPPER := 9.0;
   WRITELN;
   SIMPS(LOWER, UPPER, TOL, SUM);
   WRITELN;
   WRITELN(' area = ', SUM)
END.
```

Figure 9.8: Simpson's Rule (cont.)

Compare Figure 9.9 with Figures 9.5 and 9.7. At each point, the value of the integral obtained by Simpson's rule is closer to the correct answer than is the value obtained from the straight trapezoidal approach for the same number of panels. As a result, fewer operations are needed and so convergence occurs more quickly. On the other hand, the convergence is about the same as the trapezoidal method with end correction.

At each step of Simpson's method, the function is evaluated only at the odd positions. The sum for the even positions is obtained from the sum of all of the previous interior positions, both even and odd.

Second Run of the Simpson Program—An Exponential Function

Let us consider a second example of integration by the Simpson method. The function we will study is related to the normal distribution described in Chapter 2. It is also related to the error function described in Chapter 11. At this time, we will integrate the curve from zero to a value of unity.

Make a copy of the previous program and alter it so that it will find the value of the integral:

$$\int_0^1 e^{-x^2} dx$$

Change function FX so that it looks like:

```
FUNCTION FX(X : REAL) : REAL;
BEGIN
    FX := EXP(−X * X)
END;
```

```
              2  2.54815
              4  2.27725
              8  2.21005
             16  2.19864
             32  2.19734
             64  2.19723
            128  2.19723

         area  =   2.19723
```

Figure 9.9: Integration by Simpson's Rule

Then change the integration limits for the first and second lines of the main program so that they read:

    LOWER := 0.0;
    UPPER := 1.0;

Run the new version and compare the output to Figure 9.10. Notice how rapidly the process converges for this function.

```
              2    0.747180
              4    0.746855
              8    0.746826
             16    0.746824

         area =   0.746824
```

Figure 9.10: Integration of e^{-x^2} by Simpson's Rule

Third Run of the Simpson Program—A Periodic Function

All the functions we have integrated so far exhibit curvature in the same direction throughout the interval. Let us now consider the periodic function:

$\sin^2 x$

over the interval of 0 to 4π. This function is always positive; consequently, the integral over any region should produce a positive number. But this function has a zero value at both limits, as well as at the midpoint of the interval and at the quarter points. Consequently, the first area calculated by either the trapezoidal method or the Simpson method will give a zero result. The next approximations, however, will give positive results with Simpson's method.

Make a copy of the previous program to solve this function over the limit of 0 to 4π. Add the statement:

    PI = 3.141593;

to the constant declaration at the beginning of the program. Next, alter function FX to look like:

```
    BEGIN
        FX := SQR(SIN(X))
    END;
```

Finally, change the limits on the first line of the main program to read:

LOWER : = 0.0;
UPPER : = 4.0 * PI;

Execute the program and compare the output to Figure 9.11.

```
         2  8.57179E-12
         4  7.55984E-12
         8  8.37758
        16  6.28319
        32  6.28319

      area =   6.28319
```

Figure 9.11: The Integral of sin² x over 0 to 4π

If we had programmed this routine to quit when the absolute value of the area was less than the value of TOL, then the operation would have stopped after the first try. The resulting calculation would then be about zero, rather than the correct value of 6.28319. Thus we were justified in selecting a *relative* tolerance as the convergence criterion.

We will now derive an end-correction formula for Simpson's method.

PASCAL PROGRAM:
THE SIMPSON METHOD WITH END CORRECTION

Error correction can be applied to the Simpson method, as it was with the trapezoidal rule. The principal term contains the fourth derivative of the function. In this case, however, it is better to use an approximation requiring only the first derivative. As with the trapezoidal correction, the derivative is only needed at the end points. The Simpson's rule formula with end correction now looks like:

$$\left\{7(f_0 + f_n) + 14 \sum_{\substack{j=2 \\ j \text{ even}}}^{n-2} f_j + 16 \sum_{\substack{j=1 \\ j \text{ odd}}}^{n-1} f_j + \Delta x[f'(a) - f'(b)]\right\} \frac{\Delta x}{15}$$

Running the Program with End Correction

Make a copy of the Simpson integration program shown in Figure 9.8. Add function DFX that was used in the trapezoidal method with end correction. Alter procedure SIMPS so that it looks like Figure 9.12.

```
PROCEDURE SIMPS(LOWER, UPPER, TOL : REAL;
                            VAR SUM : REAL);

(* numerical integration by Simpson's rule *)
(* function is FX, limits are LOWER and UPPER *)
(* with number of regions equal to PIECES *)
(* partition is DELTA _ X, answer is SUM *)

VAR
    I : INTEGER;
    X, DELTA _ X, EVEN _ SUM,
    ODD _ SUM, END _ SUM,
    END _ COR, SUM1 : REAL;
    PIECES : INTEGER;

BEGIN
    PIECES : = 2;
    DELTA _ X : = (UPPER − LOWER)/PIECES;
    ODD _ SUM : = FX(LOWER + DELTA _ X);
    EVEN _ SUM : = 0.0;
    END _ SUM : = FX(LOWER) + FX(UPPER);
    END _ COR : = DFX(LOWER) − DFX(UPPER);
    SUM : = (END _ SUM + 4.0 * ODD _ SUM) * DELTA _ X/3.0;
    WRITELN(PIECES:5, SUM);

    REPEAT
        PIECES : = PIECES * 2;
        SUM1 : = SUM;
        DELTA _ X : = (UPPER − LOWER)/PIECES;
        EVEN _ SUM : = EVEN _ SUM + ODD _ SUM;
        ODD _ SUM : = 0.0;
        FOR I : = 1 TO PIECES DIV 2 DO
```

Figure 9.12: Procedure SIMPS with End Correction

```
    BEGIN
        X := LOWER + DELTA_X * (2.0 * I - 1.0);
        ODD_SUM := ODD_SUM + FX(X)
    END;
    SUM := (7.0 * END_SUM + 14.0 * EVEN_SUM
            + 16.0 * ODD_SUM + END_COR * DELTA_X)
            * DELTA_X/15.0;
    WRITELN(PIECES :5, SUM)
    UNTIL ABS(SUM - SUM1) <= ABS(TOL * SUM)
END (* simps *);
```

Figure 9.12: Procedure SIMPS with End Correction (cont.)

Run the program and compare the results to Figure 9.13. Convergence here is the fastest of the four techniques we have used. As with the trapezoidal method, however, the end-correction term cannot be used if the slope approaches infinity. Furthermore, if the function has zero slope at the upper and lower limits, then the end correction term is zero.

In the next section we will consider a somewhat more complicated technique for numerical integration. We will implement this technique in a Pascal program and then run the program on the same three functions we examined in the section on Simpson's method. This will provide an opportunity for comparison of the two methods.

```
         2   2.54815
         4   2.16287
         8   2.19490
        16   2.19713
        32   2.19722
        64   2.19722

      area =   2.19722
```

Figure 9.13: Integration by Simpson's Rule with End Correction

THE ROMBERG METHOD

The Simpson method, which uses a set of second-order equations, is an improvement over the trapezoidal method in which first-order equations are used. Accordingly, we might attempt to further improve our numerical integration method by replacing the original curve with a set of cubic, or even higher-order, polynomials. But there is another approach known as the Romberg integration method. With this technique, the area is calculated by the trapezoidal method, but the errors inherent in the trapezoidal method are accounted for by using an interpolation method.

We designate the usual sequence of the trapezoidal values by the notation t_{11}, t_{21}, t_{31}, etc. They are assigned to the first column of the two-dimensional matrix T. The first level of interpolated values is designated as t_{12}, t_{22}, t_{32}, etc. They are placed into column 2 of the T matrix. We can then interpolate between the interpolated values to produce a third column designated t_{13}, t_{23}, t_{33}, etc.

$$\begin{bmatrix} t_{11} & t_{12} & t_{13} & \cdots \\ t_{21} & t_{22} & t_{23} & \cdots \\ t_{31} & t_{32} & t_{33} & \cdots \\ \cdots & \cdots & \cdots & \cdots \end{bmatrix}$$

If we continue in this way, we will find that the interpolated values rapidly converge upon the correct integral. The advantage of this method is that the function only has to be evaluated for the entries in the first column, corresponding to the regular trapezoidal-rule values. The function does not have to be evaluated to obtain the entries for the other columns. Rather, each value is obtained from a combination of the entry directly to the left and the one just below the entry to the left. For example:

$$t_{12} = \frac{4t_{21} - t_{11}}{3}$$

$$t_{22} = \frac{4t_{31} - t_{21}}{3}$$

$$t_{13} = \frac{16t_{22} - t_{12}}{15}$$

The general algorithm is:

$$T_{ij} = \frac{4^{j-1} t_{i+1, j-1} - t_{i, j-1}}{4^{j-1} - 1}$$

PASCAL PROGRAM: INTEGRATION BY THE ROMBERG METHOD

A program that can be used to perform numerical integration by the Romberg method is given in Figure 9.14. Type up the program and execute it.

```
PROGRAM ROMB1(OUTPUT);

(* integration by the Romberg method *)
(* Dec 24, 80 *)

CONST
    TOL = 1.0E−5;

VAR
    DONE : BOOLEAN;
    SUM, UPPER, LOWER : REAL;

FUNCTION FX(X : REAL) : REAL;
(* find f(x) = 1/x *)
(* watch out for x = 0 *)

BEGIN
    FX := 1.0/X
END;

PROCEDURE ROMB(LOWER, UPPER, TOL : REAL;
                            VAR ANS : REAL);

(* numerical integration by Romberg method *)

VAR
    NX : ARRAY[1..16] OF INTEGER;
    T : ARRAY[1..136] OF REAL;
    DONE, ERROR : BOOLEAN;
    PIECES, NT, I, II, N, NN,
    L, NTRA, K, M, J : INTEGER;
    DELTA_X, C, SUM, FOTOM, X : REAL;
```

Figure 9.14: The Romberg Method

```
BEGIN
  DONE := FALSE;
  ERROR := FALSE;
  PIECES := 1;
  NX[1] := 1;
  DELTA_X := (UPPER - LOWER)/PIECES;
  C := (FX(LOWER) + FX(UPPER)) * 0.5;
  T[1] := DELTA_X * C;
  N := 1;
  NN := 2;
  SUM := C;
  REPEAT
    N := N+1;
    FOTOM := 4.0;
    NX[N] := NN;
    PIECES := PIECES * 2;
    L := PIECES - 1;
    DELTA_X := (UPPER - LOWER)/PIECES;
    (* compute trapezoidal sum for 2↑(n-1) + 1 points *)
    FOR II := 1 TO (L + 1) DIV 2 DO
      BEGIN
        I := II * 2 - 1;
        X := LOWER + I * DELTA_X;
        SUM := SUM + FX(X)
      END;

    T[NN] := DELTA_X * SUM;
    WRITE(PIECES:5, T[NN]);
    NTRA := NX[N-1];
    K := N-1;
    (* compute n-th row of T array *)
    FOR M := 1 TO K DO
```

Figure 9.14: The Romberg Method (cont.)

```
          BEGIN
              J := NN + M;
              NT := NX[N−1] + M − 1;
              T[J] := (FOTOM * T[J−1] − T[NT])/(FOTOM − 1.0);
              FOTOM := FOTOM * 4.0
          END;
        WRITELN(J :4, T[J]);
        IF N > 4 THEN
          BEGIN
            IF T[NN+1] <> 0.0 THEN
                IF (ABS(T[NTRA+1] − T[NN+1]) <= ABS(T[NN+1] * TOL))
                  OR (ABS(T[NN−1] − T[J]) <= ABS(T[J] * TOL)) THEN
                      DONE := TRUE
            ELSE IF N > 15 THEN
                BEGIN
                    DONE := TRUE;
                    ERROR := TRUE
                END
          END; (* IF n > 4 *)
        NN := J + 1
      UNTIL DONE;
      ANS := T[J]
  END (* romberg *);

BEGIN (* main program *)
    LOWER := 1.0;
    UPPER := 9.0;
    WRITELN;
    ROMB(LOWER, UPPER, TOL, SUM);
    WRITELN;
    WRITELN(' area = ', SUM)
  END.
```

Figure 9.14: The Romberg Method (cont.)

First Run of the Romberg Program

The regular trapezoidal values are printed at each step, as before. In addition, the interpolated values at the right side of the matrix are also shown. The results should look like Figure 9.15.

```
         2  3.02222     3  2.54815
         4  2.46349     6  2.25919
         8  2.27341    10  2.20472
        16  2.21733    15  2.19773
        32  2.20234    21  2.19724
        64  2.19851    28  2.19723

       area =   2.19723
```

Figure 9.15: Integration by the Romberg Method

By comparing the values in Figure 9.15 to those in the previous tables, it can be seen that the Romberg method converges even more rapidly (in this case) than Simpson's method.

Second Run of the Romberg Method—An Exponential Function

Alter the Romberg method so that it calculates the integral:

$$\int_0^1 e^{-x^2} dx$$

Change function FX so that it reads:

FUNCTION FX(X : REAL) : REAL;
BEGIN
    FX := EXP(−X ∗ X)
END;

and change the limits in the first line in the main program to read:

    LOWER := 0.0;
    UPPER := 1.0;

Run the Romberg method with the new formula. The results should look like Figure 9.16. Notice that in this case, convergence is not as rapid as it was with Simpson's method.

```
          2   0.731370     3   0.747180
          4   0.742984     6   0.746834
          8   0.745866    10   0.746824
         16   0.746585    15   0.746824

        area =  0.746824
```

Figure 9.16: Integration of e^{-x^2} by the Romberg Method

Third Run of the Romberg Method—A Periodic Function

We will now return to the integral:

$$\int_0^{4\pi} \sin^2 x \, dx$$

If we attempt to solve this equation using the Romberg method, we will essentially obtain a value of zero for the first two approximations. However, if we are careful to use a relative tolerance rather than an absolute tolerance to determine convergence, we will find the correct solution after several more iterations.

Make a copy of the Romberg program. Then change the three places that were changed in the Simpson method solution of this problem. The statement:

PI = 3.141593;

is added to the CONST section near the beginning. Function FX is altered to read:

BEGIN
    FX := SQR(SIN(X))
END;

and the first and second lines of the main program become:

LOWER := 0.0;
UPPER := 4.0 * PI;

Run the program and compare the output to Figure 9.17.

Finally, in the next section, we will expand our Romberg integration program to deal with a special case—a function that approaches infinity at one of the limits to the area beneath the curve. To solve this problem we will develop a technique of adjusting the width of the panels as they approach the infinite limit until we have arrived at a sufficiently accurate value for the total area.

FUNCTIONS THAT BECOME INFINITE AT ONE LIMIT

For each of the above methods, we have used a uniform panel width throughout the desired interval. But it should be apparent that convergence is more difficult, that is, more panels are required, when the function has a greater slope. Conversely, wider panels can be used when the slope is smaller. Furthermore, a function may be infinite at one limit even though the area for the interval is finite.

Consider, for example, the integral of the reciprocal of the square root of x over the range from zero to unity:

$$\int_0^1 \frac{dx}{\sqrt{x}}$$

The exact value of the integral is found to be 2.0 by direct integration.

If we attempt to solve this problem with one of the previous integration programs we immediately run into a problem. The value of $f(a)$ at the left edge is infinite and so we must choose some other lower limit. If we choose a small lower limit such as 10^{-11}, convergence will take a very long time. But if we choose a larger value, such as 0.01, for the left limit, the result will be inaccurate.

PASCAL PROGRAM:
ADJUSTABLE PANELS FOR AN INFINITE FUNCTION

One method of dealing with this situation is to start the integration at a value somewhat larger than zero. We will initially choose the left boundary to be a value of 0.1. The integral is then evaluated over the

```
        2 9.64326e-12    3 8.57179e-12
        4 8.08070e-12    6 7.49238e-12
        8 6.28319       10 9.07793
       16 6.28319       15 6.08755
       32 6.28319       21 6.28633

    area =   6.28633
```

Figure 9.17: Integration of sin²x by the Romberg Method

range 0.1 to 1.0. To test the reasonableness of this, we should then evaluate the next region to the left, from 0.01 to 0.1. If this area is also significant, then we must add its area to the area obtained for the first regions. We then take the next region from 0.001 to 0.01 and see how large it is. In this way, we can take regions closer and closer to zero, observing the additional area as we go. The program shown in Figure 9.18 uses this approach with the Romberg method.

Running the Program

Type up the program and execute it. The program is similar to the previous one, and so it may be easier to alter a copy of Figure 9.14.

```
PROGRAM ROMB3(OUTPUT);
(* integration by the Romberg method *)
(* Jan 9, 81 *)

CONST
    TOL = 1.0E−5;

VAR
    DONE : BOOLEAN;
    SUMT, SUM, UPPER, LOWER : REAL;

FUNCTION FX(X : REAL) : REAL;

(* find f(x) = 1/sqrt(x) *)
(* watch out for x = 0 *)

BEGIN
    FX := 1.0/SQRT(X)
END;

PROCEDURE ROMB(LOWER, UPPER, TOL : REAL;
                       VAR ANS : REAL);

(* numerical integration by Romberg method *)
```

Figure 9.18: Romberg Integration with Adjustable Panels

```
VAR
    NX : ARRAY[1..16] OF INTEGER;
    T : ARRAY[1..136] OF REAL;
    DONE, ERROR : BOOLEAN;
    PIECES, NT,I, II, N, NN,
    L, NTRA, K, M, J : INTEGER;
    DELTA_X, C, SUM, FOTOM, X : REAL;


BEGIN
    DONE := FALSE;
    ERROR := FALSE;
    PIECES := 1;
    NX[1] := 1;
    DELTA_X := (UPPER - LOWER)/PIECES;
    C := (FX(LOWER) + FX(UPPER)) * 0.5;
    T[1] := DELTA_X * C;
    N := 1;
    NN := 2;
    SUM := C;


REPEAT
    N := N + 1;
    FOTOM := 4.0;
    NX[N] := NN;
    PIECES := PIECES * 2;
    L := PIECES - 1;
    DELTA_X := (UPPER - LOWER)/PIECES;
    (* compute trapezoidal sum for 2↑(n−1) + 1 points *)
    FOR II := 1 TO (L + 1) DIV 2 DO
```

Figure 9.18: Romberg Integration with Adjustable Panels (cont.)

```
        BEGIN
            I := II * 2 - 1;
            X := LOWER + I * DELTA_X;
            SUM := SUM + FX(X)
        END;

    T[NN] := DELTA_X * SUM;
    NTRA := NX[N-1];
    K := N - 1;
    (* compute n-th row of T array *)
    FOR M := 1 TO K DO
        BEGIN
            J := NN + M;
            NT := NX[N-1] + M - 1;
            T[J] := (FOTOM*T[J-1] - T[NT])/(FOTOM - 1.0);
            FOTOM := FOTOM * 4.0
        END;
        IF N > 4 THEN
        BEGIN
            IF T[NN+1] <> 0.0 THEN
                IF (ABS(T[NTRA+1] - T[NN+1]) <= ABS(T[NN+1] * TOL))
                    OR (ABS(T[NN-1] - T[J]) <= ABS(T[J] * TOL)) THEN
                        DONE := TRUE
                ELSE IF N > 15 THEN
                    BEGIN
                        DONE := TRUE;
                        ERROR := TRUE
                    END
        END; (* IF n > 4 *)
    NN := J + 1
    UNTIL DONE;
    ANS := T[J]
END (* romberg *);
```

Figure 9.18: Romberg Integration with Adjustable Panels (cont.)

```
BEGIN (* main program *)
   LOWER := 0.1;
   UPPER := 1.0;
   WRITELN;
   SUMT := 0.0;
   WRITELN(' new area   total area   lower',
              ' upper  limits');
   REPEAT
      ROMB(LOWER, UPPER, TOL, SUM);
      UPPER := LOWER;
      LOWER := 0.1 * UPPER;
      SUMT := SUMT + SUM;
      WRITELN(SUM :9:6,' ', SUMT :9:5,
              ' ',LOWER, ' ', UPPER)
   UNTIL ABS(SUM) < TOL
END.
```

Figure 9.18: Romberg Integration with Adjustable Panels (cont.)

The WRITELN statement in procedure ROMB has been removed, so that only the final value of each major area is displayed. The main program repeatedly calls procedure ROMB with limits that are closer and closer to zero. When the new area is less than the tolerance, the procedure is terminated. The results should look like Figure 9.19.

```
    new area     total area     lower      upper   limits
    1.36754      1.36754      1.00000e-2    1.00000e-1
    0.432456     1.80000      1.00000e-3    1.00000e-2
    0.136754     1.93675      1.00000e-4    1.00000e-3
    0.043246     1.98000      1.00000e-5    1.00000e-4
    0.013675     1.99368      1.00000e-6    1.00000e-5
    0.004325     1.99800      1.00000e-7    1.00000e-6
    0.001368     1.99937      1.00000e-8    1.00000e-7
    0.000432     1.99980      1.00000e-9    1.00000e-8
    0.000137     1.99994      1.00000e-10   1.00000e-9
    0.000043     1.99998      1.00000e-11   1.00000e-10
    0.000014     1.99999      1.00000e-12   1.00000e-11
    0.000004     2.00000      1.00000e-13   1.00000e-12
```

Figure 9.19: Integration of $1/\sqrt{x}$ Near Zero

If the intermediate print statements in ROMB are not removed, it will be seen that each of the above regions has been divided into 64 panels. It should be realized that the overall width of each region is one-tenth that of the region immediately to the right and that the final left limit is 10^{-13}. Thus if the entire region from 10^{-13} to 1.0 were integrated as one unit, it would require over a million million uniformly spaced panels. Such a method would take an extremely long time to obtain the correct answer.

SUMMARY

In this chapter we have developed and compared three different numerical integration programs. The first two—the trapezoidal rule and the Simpson method implementations—utilize "end correction" for refinement of the final approximation of the area under the curve. The third, more sophisticated method—Romberg's integration—uses a matrix of progressive interpolations, and does not require error correction. We tried these methods on several different functions; in addition, we used a refined Romberg integration program to compute the area under the curve of an infinite function.

All of the above integrations except the last were performed on analytic functions with uniformly spaced panels. Other techniques are required with discrete data. One approach is to fit the data to a polynomial function, using one of the techniques we discussed in earlier chapters, then integrate the resulting polynomial.

Chapter title
CHAPTER **10**

Nonlinear Curve-Fitting Equations

INTRODUCTION

In Chapters 5 and 7 we developed computer programs for finding the coefficients to various curve-fitting equations. Since we chose approximating functions with linear coefficients, the resulting equations were linear, and therefore easily solved. Sometimes, however, it is necessary to choose approximating functions with nonlinear coefficients. In this case, the calculation of the coefficients may be more difficult.

In this chapter, we will consider two different techniques for nonlinear curve fitting. In one method, we will linearize the approximating function and then find the solution to the linear form. Using this first method, we will find curve fits for two different sets of data. First, we will fit a linearized form of the so-called *rational function* to data representing the Clausing factor. Then, we will fit a linearized exponential function to data representing the diffusion of zinc in copper over a given temperature range.

Our second approach to nonlinear curve fitting will be more direct. We will see that there is no general technique for handling approximating functions that have nonlinear coefficients; however, we will investigate a specific method for exponential equations. This method involves eliminating one of the coefficients and then solving the resulting equations with Newton's method, which we studied in Chapter 8.

Let us begin, then, with our first example of the linearization method.

LINEARIZING THE RATIONAL FUNCTION

A commonly used, nonlinear approximating function is known as the *rational function*. This expression is formed from the ratio of two polynomials:

$$y = \frac{A_1 + A_3x + A_5x^2 \ldots}{1 + A_2x + A_4x^2 + A_6x^3 \ldots}$$

In this expression, x is the independent variable, y is the dependent variable and A_1, A_2, etc., are the coefficients, as usual.

The rational function is nonlinear. However, it can be linearized by the following operations. Both sides of the equation are multiplied by the denominator polynomial, to give:

$$y(1 + A_2x + A_4x^2 + A_6x^3 \ldots) = A_1 + A_3x + A_5x^2 \ldots$$

The terms of the new equation can be rearranged to give:

$$y = A_1 - A_2xy + A_3x - A_4x^2y + A_5x^2 - A_6x^3y \ldots$$

Some of the terms on the right contain the dependent variable, y, as well as the independent variable, x. But remember, both x and y are arrays of known values. It is the unknown coefficients $A_1, A_2, \ldots A_n$ that we want. All of these coefficients are now linear, and so they can be determined by methods developed in Chapters 5 and 7.

PASCAL PROGRAM:
THE CLAUSING FACTOR FITTED TO THE RATIONAL FUNCTION

A program for producing a least-squares fit to the linearized form of the rational function is given in Figure 10.1. The program can be derived from Figure 7.3 in Chapter 7.

The data in this program represent the *Clausing factor* as a function of length-to-radius (L/r) for cylindrical orifices. When molecules with a long mean free path effuse through a cylindrical orifice, some of the molecules strike the orifice walls and are returned in the opposite direction. The remaining molecules continue through to the other side of the orifice. The Clausing factor gives the fraction of those molecules entering one end of a cylindrical orifice that actually emerge from the other end. The Clausing factor ranges from zero to unity. Of course, the Clausing factor becomes smaller as the cylinder increases in length or decreases in radius. Thus, the ratio L/r would appear in a formula for the direct calculation of the Clausing factor.

```
PROGRAM FITPOL(OUTPUT);
   (* Dec 30, 80 *)


(* Pascal program to perform a *)
(* linear least—squares fit *)
(* to the ratio of 2 polynomials *)
(* Separate procedure GAUSSJ needed *)


CONST
   MAXR = 20; (* data points *)
   MAXC = 4; (* polynomial terms *)


TYPE
   ARY   = ARRAY[1..MAXR] OF REAL;
   ARYS  = ARRAY[1..MAXC] OF REAL;
   ARY2  = ARRAY[1..MAXR, 1..MAXC] OF REAL;
   ARY2S = ARRAY[1..MAXC, 1..MAXC] OF REAL;


VAR
   X, Y, Y_CALC,
   RESID          : ARY;
   COEF, SIG      : ARYS;
   NROW, NCOL   : INTEGER;
   CORREL_COEF : REAL;
```

Figure 10.1: The Clausing Factor Fitted to the Ratio of Two Polynomials

```
PROCEDURE GET _ DATA
         (VAR X      : ARY; (* independent variable *)
          VAR Y      : ARY; (* dependent variable *)
          VAR NROW : INTEGER); (* length of vectors *)


VAR
   I : INTEGER;


BEGIN
(* Clausing Factors *)
   NROW := 10;
   X[1] := 0.1; Y[1] := 0.9524;
   X[2] := 0.2; Y[2] := 0.9092;
   X[3] := 0.5; Y[3] := 0.8013;
   X[4] := 1.0; Y[4] := 0.6720;
   X[5] := 1.2; Y[5] := 0.6322;
   X[6] := 1.5; Y[6] := 0.5815;
   X[7] := 2.0; Y[7] := 0.5142;
   X[8] := 3.0; Y[8] := 0.4201;
   X[9] := 4.0; Y[9] := 0.3566;
   X[10] := 6.0; Y[10] := 0.2755
END (* procedure get _ data *);


PROCEDURE WRITE _ DATA;
(* print out the answers *)


VAR
   I: INTEGER;


BEGIN
   WRITELN;
   WRITELN;
   WRITELN('   I   X   Y   Y CALC    RESID');
```

Figure 10.1: The Clausing Factor Fitted to the Ratio of Two Polynomials (cont.)

```
    FOR I := 1 TO NROW DO
        WRITELN (I: 3, X[I]: 8: 1, Y[I]: 9: 4,
            Y_CALC[I]: 9: 4, RESID[I]: 9: 4);
    WRITELN;
    WRITELN('coefficients   errors');
    WRITELN(COEF[1]:8:5, '   ', SIG[1], '  Constant term');
    FOR I := 2 TO NCOL DO
        WRITELN
            (COEF[I]:8:5, '   ', SIG[I]); (* other terms *)
    WRITELN;
    WRITELN(' Correlation coefficient is ', CORREL_COEF: 8: 5)
END (* write_data *);


(* PROCEDURE square(x : ary2;
                    y : ary;
                VAR a : ary2s;
                VAR g : arys;
              nrow,ncol : integer);
extern; *)
(*$F SQUARE.PAS *)


(* PROCEDURE gaussj
    (VAR b      : ary2s;
         y      : arys;
      VAR coef  : arys;
         ncol   : integer;
       VAR error : boolean);
extern; *)
(*$F GAUSSJ.PAS *)

PROCEDURE LINFIT(X,         (* independent variable *)
                Y : ARY; (* dependent variable *)
          VAR Y_CALC: ARY; (* calculated dep. variable *)
          VAR RESID    : ARY; (* array of residuals *)
```

Figure 10.1: The Clausing Factor Fitted to the Ratio of Two Polynomials (cont.)

```
        VAR  COEF   : ARYS; (* coefficients *)
        VAR  SIG    : ARYS; (* errors on coefficients *)
             NROW  : INTEGER; (* length of ary *)
        VAR  NCOL   : INTEGER); (* number of terms *)


(* least—squares fit to *)
(* nrow sets of x and y pairs of points *)
(* Separate procedures needed:
   SQUARE — form square coefficient matrix
   GAUSSJ — Gauss—Jordan elimination *)

VAR
   XMATR : ARY2; (* data matrix *)
   A       : ARY2S; (* coefficient matrix *)
   G       : ARYS; (* constant vector *)
   ERROR   : BOOLEAN;
   I, J, NM : INTEGER;
   XI, YI, YC, SRS, SEE,
   SUM_Y, SUM_Y2: REAL;


BEGIN (* procedure linfit *)
   NCOL := 4 (* number of terms *);
   FOR I := 1 TO NROW DO
      BEGIN (* setup X matrix *)
         XI := X[I];
         YI := Y[I];
         XMATR[I, 1] := 1.0 (* first column *);
         XMATR[I, 2] := —XI * YI;
         XMATR[I, 3] := XI;
         XMATR[I, 4] := —SQR(XI) * YI
      END;
```

Figure 10.1: The Clausing Factor Fitted to the Ratio of Two Polynomials (cont.)

```
SQUARE(XMATR, Y, A, G, NROW, NCOL);
GAUSSJ(A, G, COEF, NCOL, ERROR);
SUM_Y := 0.0;
SUM_Y2 := 0.0;
SRS := 0.0;
FOR I := 1 TO NROW DO
   BEGIN
      XI := X[I];
      YI := Y[I];
      YC := COEF[1]
         +(−COEF[2] * YI + COEF[3] − COEF[4] * XI * YI) * XI;
      Y_CALC[I] := YC;
      RESID[I] := YC − YI;
      SRS := SRS + SQR(RESID[I]);
      SUM_Y := SUM_Y + YI;
      SUM_Y2 := SUM_Y2 + YI * YI
   END;
CORREL_COEF := SQRT(1.0 − SRS /
         (SUM_Y2 − SQR(SUM_Y)/NROW));
IF NROW = NCOL THEN NM := 1
ELSE NM := NROW − NCOL;
SEE := SQRT(SRS/NM);
FOR I := 1 TO NCOL DO (* errors on solution *)
   SIG[I] := SEE * SQRT(A[I, I])
END (* linfit *);

BEGIN (* main program *)
  GET_DATA(X, Y, NROW);
  LINFIT(X, Y, Y_CALC, RESID, COEF, SIG, NROW, NCOL);
  WRITE_DATA
END.
```

Figure 10.1: The Clausing Factor Fitted to the Ratio of Two Polynomials (cont.)

Running the Program

Type up the program and execute it. The matrix is calculated for four terms, corresponding to a first-order numerator and a second-order denominator. Additional terms can be easily added to the approximating function. The symbols MAXC and NCOL must be changed to reflect the actual number of terms. In addition, expressions such as:

> MATR[I,5] := MATR[I,3] * XI;
>
> MATR[I,6] := MATR[I,4] * XI;

must be added to procedure LINFIT if additional terms are desired. The results shown in Figure 10.2 correspond to the equation:

$$y = \frac{1.0017 + 0.237x}{1 + 0.7522x + 0.0912x^2}$$

If the data are fitted with a regular polynomial function rather than the rational function, the resulting fit will not be as good (for the same number of coefficients in the approximating function).

In our next example of the linearization approach, we will examine an exponential equation. Later in this chapter we will fit the same equation using another, more direct approach.

```
        I      X        Y       Y CALC    RESID
        1     0.1    0.9524    0.9529    0.0005
        2     0.2    0.9092    0.9090   -0.0002
        3     0.5    0.8013    0.8006   -0.0007
        4     1.0    0.6720    0.6719   -0.0001
        5     1.2    0.6322    0.6324    0.0002
        6     1.5    0.5815    0.5818    0.0003
        7     2.0    0.5142    0.5146    0.0004
        8     3.0    0.4201    0.4199   -0.0002
        9     4.0    0.3566    0.3563   -0.0003
       10     6.0    0.2755    0.2756    0.0001

    coefficients      errors
     1.00169        4.60364e-4   Constant term
     0.75220        1.37135e-2
     0.23704        1.22040e-2
     0.09124        5.14729e-3

    Correlation coefficient is   1.00000
```

Figure 10.2: The Clausing Factor vs L/r Fitted to a Rational Function

LINEARIZING THE EXPONENTIAL EQUATION

One of the most common nonlinear equations has the form:

$$y = Ae^{Bx}$$

where x is the independent variable, y is the dependent variable, and A and B are the desired coefficients. This equation occurs widely throughout science and engineering because it is the solution to a first-order differential equation.

This equation can be linearized by taking the logarithm. The result is:

$$\ln y = \ln A + Bx$$

In this form, the dependent variable is $\ln y$ and the unknown coefficients are $\ln A$ and B. Since the new coefficients are linear, a least-squares fit can be obtained with procedure LINFIT that was developed in Chapter 5.

PASCAL PROGRAM:
AN EXPONENTIAL CURVE FIT FOR THE DIFFUSION OF ZINC IN COPPER

Figure 10.3 gives a program for finding a least-squares fit to the linearized exponential equation. The program can be derived from Figure 7.3 in Chapter 7. The value of NCOL has been changed to 2, and the calculation of the standard error has been removed.

The data given in procedure GET_DATA represent the diffusion of zinc in copper over the temperature range 600° to 900° C. The diffusion equation is:

$$D = D_0 e^{-Q/RT}$$

where D is the diffusion coefficient in cm sq/sec, D_0 is the diffusion constant in the same units, Q is the activation energy in cal/mole, R is the gas constant, and T is the temperature in degrees Kelvin. The independent variable, x, is the reciprocal of the temperature in degrees Kelvin. The dependent variable, y, is the logarithm of the diffusion coefficient.

```pascal
PROGRAM DIFFUS(OUTPUT);
   (* Dec 30, 80 *)
(* Pascal program to perform a *)
(* linear least—squares fit *)
(* for the diffusion of Zn in Cu *)
(* External procedure GAUSSJ needed *)
CONST
   MAXR = 20 (* data points *);
   MAXC = 4 (* polynomial terms *);
   R    = 1.987 (* gas constant *);
TYPE
   ARY   = ARRAY[1..MAXR] OF REAL;
   ARYS  = ARRAY[1..MAXC] OF REAL;
   ARY2  = ARRAY[1..MAXR, 1..MAXC] OF REAL;
   ARY2S = ARRAY[1..MAXC, 1..MAXC] OF REAL;
VAR
   X, Y, Y_CALC, T, D,
   RESID        : ARY;
   COEF, SIG    : ARYS;
   NROW, NCOL   : INTEGER;
   CORREL_COEF, SRS : REAL;
PROCEDURE GET_DATA(VAR X, Y, T, D : ARY;
                   VAR NROW     : INTEGER);
(* get values for nrow and arrays t, d *)
VAR
   I : INTEGER;
BEGIN
   NROW := 7;
   T[1] := 600.0; D[1] := 1.4E−12;
   T[2] := 650.0; D[2] := 5.5E−12;
   T[3] := 700.0; D[3] := 1.8E−11;
   T[4] := 750.0; D[4] := 6.1E−11;
   T[5] := 800.0; D[5] := 1.6E−10;
```

Figure 10.3: A Least-Squares Fit to the Linearized Exponential Equation

```
    T[6] := 850.0; D[6] := 4.4E-10;
    T[7] := 900.0; D[7] := 1.2E-9;
    FOR I := 1 TO NROW DO
       BEGIN
          X[I] := 1.0/(T[I] + 273.0);
          Y[I] := LN(D[I])
       END
END (* procedure get_data *);

PROCEDURE WRITE_DATA;
(* print out the answers *)

VAR
   I: INTEGER;
BEGIN
   WRITELN;
   WRITELN;
   WRITELN('   I   T C   D   ',
          '     D Calc');
   FOR I := 1 TO NROW DO
      WRITELN(I:3, T[I]:8:0, D[I], ' ',
      Y_CALC[I]);
   WRITELN;
   WRITELN('coefficients');
   WRITELN(COEF[1], '     Constant term');
   FOR I := 2 TO NCOL DO
      WRITELN
      (COEF[I]); (* other terms *)
   WRITELN;
   WRITELN(' D0 = ', (EXP(COEF[1])) :7:2,' cm sq/sec.');
   WRITELN(' Q = ', (-R * COEF[2]/1000.0) :8:2,
             ' kcal/mole');
   WRITELN; WRITELN(' SRS = ', SRS :7:3)
END (* write_data *);
```

Figure 10.3: A Least-Squares Fit to the Linearized Exponential Equation (cont.)

```
(* PROCEDURE square(x : ary2;
                      y: ary;
                 VAR a : ary2s;
                 VAR g : arys;
            nrow,ncol : integer);
extern; *)
(*$F SQUARE.PAS *)

(* PROCEDURE gaussj
   (VAR b    : ary2s;
        y    : arys;
    VAR coef : arys;
        ncol : integer;
    VAR error : boolean);
extern; *)
(*$F GAUSSJ.PAS *)

PROCEDURE LINFIT(X,         (* independent variable *)
                   Y : ARY; (* dependent variable *)
       VAR Y_CALC: ARY; (* calculated dep. variable *)
       VAR RESID    : ARY; (* array of residuals *)
       VAR  COEF    : ARYS; (* coefficients *)
       VAR  SIG     : ARYS; (* errors on coefficients *)
            NROW  : INTEGER; (* length of ary *)
       VAR  NCOL   : INTEGER); (* number of terms *)

(* least—squares fit to *)
(* nrow sets of x and y pairs of points *)
(* Separate procedures needed:
   SQUARE — form square coefficient matrix
   GAUSSJ — Gauss—Jordan elimination *)
```

Figure 10.3: A Least-Squares Fit to the Linearized Exponential Equation (cont.)

```
VAR
    XMATR  : ARY2 (* data matrix *);
    A      : ARY2S (* coefficient matrix *);
    G      : ARYS (* constant vector *);
    ERROR  : BOOLEAN;
    I, J, NM: INTEGER;
    SEE, A1 : REAL;


BEGIN (* procedure linfit *)
    NCOL := 2 (* number of terms *);
    FOR I := 1 TO NROW DO
        BEGIN (* set up X matrix *)
            XMATR[I, 1] := 1.0 (* first column *);
            XMATR[I, 2] := X[I] (* second column *)
        END;
    SQUARE(XMATR, Y, A, G, NROW, NCOL);
    GAUSSJ(A, G, COEF, NCOL, ERROR);
    SRS := 0.0;
    A1 := EXP(COEF[1]);
    FOR I := 1 TO NROW DO
        BEGIN
            Y_CALC[I] := A1 * EXP(COEF[2] * X[I]);
            IF Y[I] <> 0.0 THEN RESID[I] := Y_CALC[I]/Y[I] − 1.0
            ELSE RESID[I] := Y[I]/Y_CALC[I] − 1.0;
            SRS := SRS + SQR(RESID[I])
        END
END (* linfit *);


BEGIN (* main program *)
    GET_DATA(X, Y, T, D, NROW);
    LINFIT(X, Y, Y_CALC, RESID, COEF, SIG, NROW, NCOL);
    WRITE_DATA
END.
```

Figure 10.3: A Least-Squares Fit to the Linearized Exponential Equation (cont.)

Running the Program

Compile the program, execute it, and compare the results with Figure 10.4. The diffusion constant (D0) is calculated from the anti-logarithm (the exponent) of the first coefficient:

D0 := EXP(COEF[1])

Multiplication of the second coefficient (COEF[2]) by the gas constant and changing the sign gives the activation energy Q:

Q := −R * COEF[2]

In this example, the sum of residuals squared, *SRS*, is given rather than the usual standard error on the coefficients. The nonlinear transform of the approximating function makes these sigmas meaningless. The calculation of this form of *SRS* is discussed more fully in the next section, where we will develop our second approach to nonlinear curve fitting.

```
  I     T C        D            D Calc
  1     600.  1.40000e-12    1.31283e-12
  2     650.  5.50000e-12    5.43316e-12
  3     700.  1.80000e-11    1.94311e-11
  4     750.  6.10000e-11    6.13542e-11
  5     800.  1.60000e-10    1.74041e-10
  6     850.  4.40000e-10    4.49924e-10
  7     900.  1.20000e-9     1.07266e-9

  Coefficients
  -1.13952         Constant term
  -2.28895e4

  D0 =      0.32 cm sq/sec.
  Q  =     45.48 kcal/mole

  SRS =   7.000
```

Figure 10.4: The Diffusion of Zinc in Copper (Linearized Fit)

DIRECT SOLUTION OF THE EXPONENTIAL EQUATION

In the previous section, the exponential equation:

$$y = Ae^{Bx}$$

was linearized by taking the logarithm:

$$\ln y = \ln A + Bx$$

A least-squares fit was then made to this linearized form of the equation. But the coefficients that produce the minimum sum of residuals squared (SRS) to the linearized form will not, in general, produce the minimum SRS for the original, nonlinearized equation. Thus we should consider a direct, least-squares solution of the nonlinear equation.

In this section, we will derive the curve-fitting equations with respect to the original nonlinearized exponential equation. If we approach the solution to the nonlinear equation the way we approached the linear equation, we will obtain the SRS for the approximating function. Then the derivative of SRS with respect to each coefficient is set to zero. There is one equation for each unknown. Now, however, the resulting equations are nonlinear and therefore cannot generally be solved.

While there is no universal approach to the solution of nonlinear equations, there are several techniques that can be used for special forms. In the case of the exponential function, the solution is relatively easy. We will follow the usual curve-fit algorithm up to the point of taking the derivatives of the SRS. Then we will see a way to eliminate one of the coefficients from the equation, and we will solve the resulting function using Newton's method.

Calculating the SRS

The residuals for the equation:

$$y = Ae^{Bx}$$

could be defined as:

$$r = Ae^{Bx} - y$$

However, if the data are all measured to about the same relative degree of precision, independent of the magnitude, it will be more meaningful to use a relative residual:

$$r = (Ae^{Bx} - y) / y$$

or

$$r = (Ae^{Bx} / y) - 1$$

The residuals in the new form are squared, and then summed to form the SRS. The derivative of the resulting SRS is taken with respect to both

A and B, and the resulting equations are set to zero. This approach is the same as before. There are two equations and two unknowns:

$$A\Sigma(e^{2Bx} / y^2) - \Sigma(e^{Bx} / y) = 0$$

and

$$A\Sigma(xe^{2Bx} / y^2) - \Sigma(xe^{Bx} / y) = 0$$

Now, however, the resulting equations are nonlinear and so they cannot be solved with procedure LINFIT.

Eliminating Coefficient A

Fortunately, the coefficient A is linear in this case, and so it can be separated. The first of the two above equations can be rearranged to give:

$$A = \Sigma(e^{Bx} / y) / \Sigma(e^{2Bx} / y^2)$$

Then, this expression for A is substituted into the second equation to give:

$$\Sigma(e^{Bx} / y) \, \Sigma(xe^{2Bx} / y^2)$$
$$- \Sigma(e^{2Bx} / y^2) \, \Sigma(xe^{Bx} / y) = 0 = f(B)$$

Since we have eliminated A, this equation is only a function of B. In Chapter 8 we developed a program to solve nonlinear equations by Newton's method. The approach is applicable here.

Applying Newton's Method

We next take the derivative of the above equation with respect to B. The result is:

$$f'(B) = 2\Sigma(e^{Bx} / y) \, \Sigma(x^2 e^{2Bx} / y^2)$$
$$- \Sigma(xe^{2Bx} / y^2) \, \Sigma(xe^{Bx} / y)$$
$$- \Sigma(e^{2Bx} / y^2) \, \Sigma(x^2 e^{Bx} / y)$$

Before going any further, check to see that all of the terms in $f(B)$ and $f'(B)$ are homogeneous. The units of each term of $f(B)$ must correspond to:

$$\frac{x}{y^3}$$

and the units of $f'(B)$ must correspond to

$$\frac{x^2}{y^3}$$

PASCAL PROGRAM:
A NONLINEARIZED EXPONENTIAL CURVE FIT

The program shown in Figure 10.5 can be used to find a nonlinear least-squares curve fit to the diffusion equation:

$$D = D_0 e^{-Q/RT}$$

This is the equation we considered earlier in the chapter. Since we are using Newton's method, we need a first approximation for the value of B. We can obtain a good first value from the linear equation for B. But instead of performing the complete linearized fit, we can simply calculate the value of B from Equation 20 of Chapter 5. In this case, however, the value of y is replaced by $\ln y$.

$$B = \frac{\Sigma[x \ln (y)] - \Sigma(x) \Sigma[\ln (y) / n]}{\Sigma x^2 - (\Sigma x)^2 / n}$$

The instructions for this first approximation are included in procedure NLIN. A more complicated approach would be to call the Gauss-Jordan procedure (Figure 4.5) for the first approximation.

Running the Program

Type up the program and execute it. Compare the results to Figure 10.6.

```
PROGRAM NLIN3(OUTPUT);
   (* Jan 23, 81 *)

(* Pascal program to perform a *)
(* nonlinear least—squares fit *)
(* for the diffusion of Zn in Cu *)

CONST
   MAXR = 20 (* data points *);
   MAXC = 4 (* polynomial terms *);
   R    = 1.987 (* gas constant *);

TYPE
   INDEX = 1..MAXR;
   ARY   = ARRAY[INDEX] OF REAL;
   ARYS = ARRAY[1..MAXC] OF REAL;
   ARY2 = ARRAY[1..MAXR, 1..MAXC] OF REAL;

VAR
   X, Y, Y_CALC, T, D, EX : ARY;
   COEF          : ARYS (* solution vector *);
   I, N          : INTEGER;
   NROW, NCOL : INTEGER;
   DONE, ERROR : BOOLEAN;
   CORREL_COEF, A, B, X2, SRS : REAL;

PROCEDURE GET_DATA(VAR X, Y : ARY;
                       VAR N: INTEGER);

(* get values for n and arrays t, d *)

VAR
   I : INTEGER;
```

Figure 10.5: A Least-Squares Fit to the Nonlinearized Exponential Function

```
BEGIN
   N := 7;
   T[1] := 600.0; D[1] := 1.4E−12;
   T[2] := 650.0; D[2] := 5.5E−12;
   T[3] := 700.0; D[3] := 1.8E−11;
   T[4] := 750.0; D[4] := 6.1E−11;
   T[5] := 800.0; D[5] := 1.6E−10;
   T[6] := 850.0; D[6] := 4.4E−10;
   T[7] := 900.0; D[7] := 1.2E−9;
   FOR I := 1 TO N DO
      BEGIN
         X[I] := 1.0/(T[I] + 273);
         Y[I] := D[I]
      END
END (* procedure get_data *);


PROCEDURE WRITE_DATA;


(* print out the answers *)


VAR
   I : INTEGER;

BEGIN
   WRITELN;
   WRITELN('  I   T C   D ',
           '   D Calc');
   FOR I := 1 TO N DO
      WRITELN(I:3, T[I]:8:0, D[I],
              '  ', Y_CALC[I]);
   WRITELN;
   WRITELN(' Coefficients');
   WRITELN(COEF[1], '    Constant term');
```

Figure 10.5: A Least-Squares Fit to the Nonlinearized Exponential Function (cont.)

```
    FOR I := 2 TO NCOL DO
        WRITELN(COEF[I]) (* other terms *);
    WRITELN;
    WRITELN(' D0 = ', A:7:2,' cm sq/sec.');
    WRITELN(' Q = ', −R * B/1000 :8:2, ' kcal/mole');
    WRITELN; WRITELN(' SRS = ', SRS :8:4)
END (* write _ data *);

PROCEDURE FUNC(B : REAL;
        VAR FB, DFB : REAL);

VAR
    I : INTEGER;
    S1, S2, S3, S4, S5, S6,
    EX1, EX2, XI, X2, YI, Y2 : REAL;

BEGIN
    S1 := 0.0;
    S2 := 0.0;
    S3 := 0.0;
    S4 := 0.0;
    S5 := 0.0;
    S6 := 0.0;
    FOR I := 1 TO N DO
      BEGIN
        XI := X[I];
        X2 := XI * XI;
        YI := Y[I];
        Y2 := YI * YI;
        EX1 := EXP(B * XI);
        EX[I] := EX1;
        EX2 := EX1 * EX1;
        S1 := S1 + XI * EX2/Y2;
```

Figure 10.5: A Least-Squares Fit to the Nonlinearized Exponential Function (cont.)

```
          S2 := S2 + EX1/YI;
          S3 := S3 + XI * EX1/YI;
          S4 := S4 + EX2/Y2;
          S5 := S5 + 2.0 * X2 * EX2/Y2;
          S6 := S6 + X2 * EX1/YI
      END;
    FB := S1 * S2 - S3 * S4;
    DFB := S2 * S5 - S1 * S3 - S4 * S6;
    A := S2/S4
END (* func *);

PROCEDURE NEWTON
          (VAR X: REAL);
  (* Jan 23, 81 *)
CONST
    TOL = 1.0E-6;
    MAX = 20;
VAR
    FX, DFX : REAL;
    DX, X1 : REAL;
    I : INTEGER;

BEGIN (* newton *)
    ERROR := FALSE;
    I := 0;
    REPEAT
      I := I + 1;
      X1 := X;
      FUNC(X, FX, DFX);
      IF DFX = 0.0 THEN
        BEGIN
          ERROR := TRUE;
          X := 1.0;
          WRITELN('ERROR: slope zero')
        END
```

Figure 10.5: A Least-Squares Fit to the Nonlinearized Exponential Function (cont.)

```
            ELSE
               BEGIN
                  DX := FX/DFX;
                  X := X1 − DX;
               END
         UNTIL
            ERROR OR
            (I > MAX) OR
            (ABS(DX) <= ABS(TOL * X));
         IF I > MAX THEN
            BEGIN
               WRITELN ('ERROR: no convergence in ',
                     MAX, ' loops');
               ERROR := TRUE
            END
   END (* newton *);
   PROCEDURE NLIN(X, Y : ARY;
            VAR Y _ CALC : ARY;
                     N : INTEGER);
   (* fit the diffusion equation through
      n sets of x and y pairs of points *)
   VAR
      RESID : ARY;
      MATR : ARY2;
      I     : INTEGER;
      XI, YI, SUM _ X,
      SUM _ Y, SUM _ Y2, B1,
      SUM _ XY, SUM _ X2 : REAL;
   BEGIN (* nlin *)
      NCOL := 2 (* two terms *);
      SUM _ X := 0.0;
      SUM _ Y := 0.0;
      SUM _ XY := 0.0;
```

Figure 10.5: A Least-Squares Fit to the Nonlinearized Exponential Function (cont.)

```
        SUM_X2 := 0.0;
        FOR I := 1 TO N DO
          BEGIN
             XI := X[I];
             YI := LN(Y[I]);
             SUM_X := SUM_X + XI;
             SUM_Y := SUM_Y + YI;
             SUM_Y2 := SUM_Y2 + YI * YI;
             SUM_XY := SUM_XY + XI * YI;
             SUM_X2 := SUM_X2 + XI * XI
          END;

        B := (SUM_XY − SUM_X * SUM_Y/N) /
             (SUM_X2 − SQR(SUM_X)/N);
        NEWTON(B);
        COEF[1] := A;
        COEF[2] := B;
        SRS := 0.0;
        FOR I := 1 TO N DO
          BEGIN
             Y_CALC[I] := A * EX[I];
             IF Y[I] <> 0.0 THEN
                RESID[I] := Y_CALC[I]/Y[I] − 1.0
             ELSE RESID[I] := Y[I]/Y_CALC[I] − 1.0;
             SRS := SRS + SQR(RESID[I])
          END
      END (* nlin *);

BEGIN (* main program *)
   GET_DATA (X, Y, N);
   NLIN(X, Y, Y_CALC, N);
   WRITE_DATA
END.
```

Figure 10.5: A Least-Squares Fit to the Nonlinearized Exponential Function (cont.)

```
     I      T C        D                D Calc
     1      600. 1.40000e-12      1.31051e-12
     2      650. 5.50000e-12      5.41377e-12
     3      700. 1.80000e-11      1.93304e-11
     4      750. 6.10000e-11      6.09471e-11
     5      800. 1.60000e-10      1.72657e-10
     6      850. 4.40000e-10      4.45808e-10
     7      900. 1.20000e-9       1.06167e-9

     Coefficients
     3.08933e-1          Constant term
    -2.28603e4

     DO =      0.31 cm sq/sec.
     Q  =     45.42 kcal/mole

     SRS =   0.0295
```

Figure 10.6: The Diffusion of Zinc in Copper (Nonlinear Fit)

Now, compare the results of Figure 10.4 to Figure 10.6. The values for the diffusion constant, D_0, and the activation energy, Q, are about the same. However, the sum of residuals squared, *SRS*, is smaller for the nonlinearized fit. It should be pointed out, however, that if the linearized residuals:

$$r = \ln D_0 - \frac{Q}{RT}$$

are used in *SRS*, the linearized *SRS* will be smaller. Furthermore, both linearized and nonlinearized programs will give exactly the same results if the data precisely follow an exponential equation. Therefore, it might be wise to perform both the linearized and nonlinearized curve fits. If the results are very different, then an error in measuring or in recording the data should be suspected.

SUMMARY

We have seen examples of two approaches to nonlinear curve fitting. The forms of the equations we used in our examples were (1) the rational function, and (2) the exponential function. We used a linearization approach and a direct approach. It is important to keep in mind that neither of the approaches we have studied represents a *general* method for nonlinear curve fitting; rather we have examined specific techniques that can be used on specific curve-fitting equations.

CHAPTER **11**

Advanced Applications:

The Normal Curve, the Gaussian Error Function, the Gamma Function, and the Bessel Functions

INTRODUCTION

This chapter takes up several advanced topics in programming for mathematical applications. The programs use a number of tools we have developed in this book. We will study three functions, all of which have important applications in mathematics, physics, and engineering: the Gaussian error function, the Gamma function, and the Bessel functions.

For evaluating the Gaussian error function we will write two programs; the first uses Simpson's rule for numeric integration, and the second uses an infinite series expansion. In our discussion of the Gaussian error function we will again take up the topic of diffusion, which we studied in Chapter 10 in the context of nonlinear curve fitting.

We will then examine the Gamma function. Following a study of the special properties of the function, we will develop a program to evaluate it. We will subsequently use this program in the last topic of the chapter—an investigation of numerical solutions to the Bessel equation. We will consider Bessel functions of the first and second kind.

The precision and range of a Pascal compiler will be significant issues in running the programs of this chapter. Thus, it will be important to recall the results of the evaluation programs we developed and ran in Chapter 1.

Let us begin by reviewing the concepts of the distribution functions.

THE NORMAL AND CUMULATIVE DISTRIBUTION FUNCTIONS

We saw in Chapter 2 that random errors, introduced during experimental measurement, cause a sequence of observed values to be dispersed about the mean or average. A frequency plot of the resulting data shows a bell-shaped curve. This shape is described as a normal distribution or a probability density function, as in Figure 11.1.

The normal distribution is defined by the equation:

$$f(x) = \frac{e^{\frac{-x^2}{2}}}{\sqrt{2\pi}} \tag{1}$$

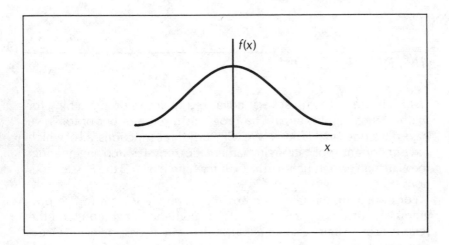

Figure 11.1: The Normal Probability Distribution Function

This function has a peak, or mean value, at $x = 0$, and ranges from minus infinity to plus infinity. The entire area under the probability-density curve (above the x-axis) is normalized to a value of unity. That is,

$$\int_{-\infty}^{\infty} f(x)dx = 1 \tag{2}$$

From the symmetry of the curve, it can be seen that the area from $x = 0$ to infinity, that is, the right half of the curve, is equal to one-half the total area.

The area from $x = -\infty$ to $x = b$ is called the *cumulative distribution function F(x)*.

$$F(x) = \int_{-\infty}^{b} f(x)dx \tag{3}$$

This integral cannot be solved in closed form, but it is tabulated in handbooks. Sometimes the integral:

$$G(x) = \int_{0}^{b} f(x)dx \tag{4}$$

is given instead. Because the curve is normalized, either integral can readily be obtained from the other. The relationship is:

$$F(x) = G(x) + 0.5$$

In the next section we will consider some numerical methods for obtaining the area $G(x)$. But first, let us further explore the normal curve.

The Standard Deviation

The area under the normal curve of the measured values is related to the standard deviation. A small standard deviation corresponds to a close grouping of the measured values about the mean. Conversely, a large standard deviation corresponds to values that are spread further from the mean. Two normal distributions are shown in Figure 11.2. Both have the same mean value, but they have different standard deviations. The curve with the smaller standard deviation has the higher peak at zero. However, both have the same area underneath the curve.

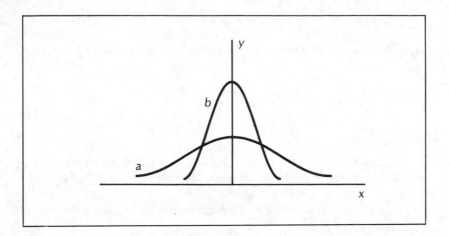

Figure 11.2: Normal Curves with Large (a) and Small (b) Standard Deviations

Figure 11.3 shows a plot of the distribution function described by Equation 3. This function has a value of 0.5 at $x = 0$ and rises asymptotically to a value of unity as x approaches infinity.

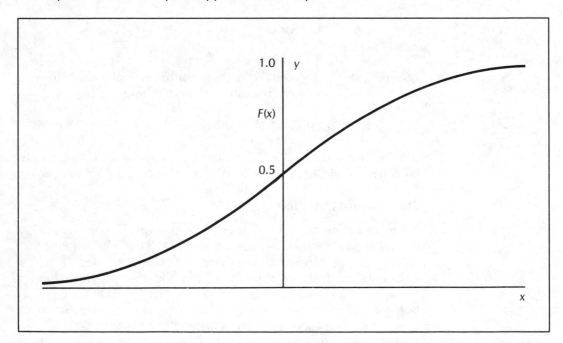

Figure 11.3: The Cumulative Distribution Function

The function $G(x)$ can be used to find the relationship between the standard deviation and the corresponding fraction of a particular sample. For example, $G(1)$ has a value of 0.34. Thus, 34% of the population lies in the range of zero to one standard deviation. Twice that value, 68%, corresponds to a range of one sigma on both sides of the mean. The distribution function can be readily obtained from the Gaussian error function, which we will discuss in the next section.

THE GAUSSIAN ERROR FUNCTION

Before formulating the Gaussian error function and implementing a program to evaluate it, we will digress slightly to a familiar application area—*diffusion*. The equation describing the diffusion of one kind of atom into another will finally lead us back to our Gaussian error function.

Diffusion in a One-Dimensional Slab

Diffusion is the net flow of atoms, electrons, or heat from a more concentrated region to a less concentrated region. The resulting flux, J, can be described by *Fick's first law*:

$$J = -D \, \frac{dC}{dx}$$

where C is the concentration, and x is the distance. The *diffusion coefficient* or *diffusivity*, D, is the same quantity we used to describe the diffusion of zinc in copper in Chapter 10. The minus sign reflects the fact that the flux occurs in a direction that is opposite to the concentration gradient. If we are concerned with the flow of atoms, then J is expressed in units of atoms per unit area-sec. The concentration, C, is given in units of atoms per unit volume.

Fick's first law can also be used to describe the flow of electrons in a conductor. We then write:

$$J = \sigma \frac{dV}{dx}$$

In this case, J is the current density, V is the voltage, and x is the distance. Sigma is the electrical conductivity.

In a similar way, the flow of heat can be expressed as:

$$J = k \frac{dT}{dx}$$

where J is the thermal conductivity in units of energy/area-sec, and T is the temperature.

Fick's second law:

$$\frac{\partial C}{\partial t} = D \frac{\partial^2 C}{\partial x^2}$$

can be used to describe the concentration as a function of position and time. There are many possible solutions to Fick's second law, depending on the boundary conditions. For example, steady-state conditions occur when the concentration no longer changes with time. Since

$$\frac{\partial C}{\partial t} = 0$$

then

$$\frac{D \partial^2 C}{\partial x^2} = 0$$

This implies that a concentration gradient, $\partial C / \partial x$ is uncurving or straight.

Another useful solution to Fick's second law describes the diffusion of one kind of atom into another. The surface concentration of the diffusing species is kept constant for all time. The other species is in the shape of a one-dimensional, semi-infinite slab. In this case, the solution to Fick's law is:

$$\frac{C_x - C_0}{C_s - C_0} = 1 - \text{erf}(y) = 1 - \text{erf}\left(\frac{x}{2\sqrt{Dt}}\right)$$

In this expression, C_x is the concentration of the diffusing species at time t and a distance x from the surface. C_s is the surface concentration that is constant for all time, and C_0 is the initial, uniform concentration for all x at time equal to zero. D is the diffusion coefficient and y has the value:

$$\frac{x}{2\sqrt{Dt}}$$

If the initial concentration, C_0, is zero, then the equation reduces to:

$$\frac{C_x}{C_s} = 1 - \text{erf}(y) = 1 - \text{erf}\left(\frac{x}{2\sqrt{Dt}}\right)$$

The quantity erf is the Gaussian error function. It is defined as:

$$\text{erf}(y) = \frac{2}{\sqrt{\pi}} \int_0^y e^{-t^2} dt \tag{5}$$

The functions $F(x)$, given in Equation 3, and $G(x)$, given in Equation 4, can be obtained from the error function by the relationship:

$$F(x) = \frac{1 + \text{erf}\left(\frac{x}{\sqrt{2}}\right)}{2}$$

$$G(x) = \frac{\text{erf}\left(\frac{x}{\sqrt{2}}\right)}{2}$$

For example, the range represented by two standard deviations on either side of the mean can be found by the error function:

$$2G(2) = \text{erf}\left(\frac{2}{\sqrt{2}}\right) = 95\%$$

PASCAL PROGRAM:
EVALUATING THE GAUSSIAN ERROR FUNCTION USING SIMPSON'S RULE

The error function cannot be solved in closed form. However, it is possible to obtain particular solutions. A straightforward approach is to use a numerical integration technique such as Simpson's rule. Figure 11.4 gives a program for finding the error function in this way.

```
PROGRAM ERFSIMP(INPUT, OUTPUT);

(* integration by Simpson's method *)
(* Feb 9, 81 *)

CONST
    TOL = 1.0E−5;
    PI  = 3.141593;

VAR
    DONE : BOOLEAN;
    SUM, UPPER, LOWER, ERF,
    TWOPI : REAL;
```

Figure 11.4: The Gaussian Error Function by Simpson's Rule

```
FUNCTION FX(X : REAL) : REAL;

BEGIN
    FX := EXP(−X * X)
END; (* function fx *)

PROCEDURE SIMPS(LOWER, UPPER, TOL :  REAL;
                              VAR SUM : REAL);

(* numerical integration by Simpson's rule *)
(* function is fx, limits are lower and upper *)
(* with number of regions equal to pieces *)
(* partition is delta_x, answer is sum *)

VAR
    I : INTEGER;
    X, DELTA_X, EVEN_SUM,
    ODD_SUM, END_SUM, SUM1 : REAL;
    PIECES : INTEGER;

BEGIN
    PIECES := 2;
    DELTA_X := (UPPER − LOWER)/PIECES;
    ODD_SUM := FX(LOWER + DELTA_X);
    EVEN_SUM := 0.0;
    END_SUM := FX(LOWER) + FX(UPPER);
    SUM := (END_SUM + 4.0 * ODD_SUM) * DELTA_X/3.0;
    REPEAT
        PIECES := PIECES * 2;
        SUM1 := SUM;
        DELTA_X := (UPPER − LOWER)/PIECES;
        EVEN_SUM := EVEN_SUM + ODD_SUM;
        ODD_SUM := 0.0;
        FOR I := 1 TO PIECES DIV 2 DO
```

Figure 11.4: The Gaussian Error Function by Simpson's Rule (cont.)

```
          BEGIN
              X := LOWER + DELTA_X * (2.0 * I − 1.0);
              ODD_SUM := ODD_SUM + FX(X)
          END;
       SUM := (END_SUM + 4.0 * ODD_SUM
              + 2.0 * EVEN_SUM) * DELTA_X/3.0
     UNTIL ABS(SUM1 − SUM) <= ABS(TOL * SUM1)
  END (* simps *);


  BEGIN (* main program *)
     DONE := FALSE;
     TWOPI := 2.0/SQRT(PI);
     LOWER := 0.0;


     REPEAT
        WRITELN;
        WRITE(' Erf? ');
        READLN(UPPER);
        IF UPPER < 0.0 THEN DONE := TRUE
        ELSE IF UPPER = 0.0 THEN
           WRITELN(' Erf of 0.0 is 0.0')


        ELSE (* upper > 0 *)
           BEGIN
              SIMPS(LOWER, UPPER, TOL, SUM);
              ERF := TWOPI * SUM;
              WRITELN(' Erf of ', UPPER:7:2,
                 ', is ', ERF:8:4)
           END
     UNTIL DONE
  END.
```

Figure 11.4: The Gaussian Error Function by Simpson's Rule (cont.)

Running the Program

The program repeatedly cycles, asking the user for input. The error function is calculated by the Simpson method, then the argument and the corresponding error function are printed. The program can be terminated by entering a negative value.

There are several disadvantages to calculating the error function by this method. The execution time increases and the accuracy decreases as the argument increases. For a six or seven digit, floating-point package, the useful range of arguments is from zero to 3.

PASCAL PROGRAM:
EVALUATING THE GAUSSIAN ERROR FUNCTION USING AN INFINITE SERIES EXPANSION

Another way to evaluate the error function is to substitute an infinite series. The new expression is then integrated term by term to produce another infinite series. The result is:

$$\text{erf}(y) = \frac{2}{\sqrt{\pi}} \; e^{-y^2} \sum_{n=0}^{\infty} \frac{2^n y^{2n+1}}{1\cdot 3\cdot \ldots \cdot(2n+1)}$$

Figure 11.5 gives a program for evaluating the error function in this way. Type up the program, compile it, and execute it. The user is asked to input an argument to the error function. Then the argument and the resulting function are printed.

The infinite series is evaluated in procedure ERF. Each new term is added to the sum. If a particular term does not change the sum by more than the value of TOL, then the routine is terminated and the current value is returned.

```
PROGRAM ERFD(INPUT, OUTPUT);

(* evaluation of the Gaussian error function *)
(* Jan 23, 81 *)

VAR
    X, ANS : REAL;
    DONE : BOOLEAN;
```

Figure 11.5: An Infinite Series Expansion for the Gaussian Error Function

```
FUNCTION ERF(X : REAL) : REAL;

(* Infinite series expansion of the
   Gaussian error function *)
CONST
   SQRTPI = 1.7724538;
   TOL = 1.0E−6;

VAR
   X2, SUM, SUM1, TERM : REAL;
   I : INTEGER;

BEGIN
   IF X = 0.0 THEN ERF := 0.0
   ELSE IF X > 4.0 THEN ERF := 1.0
   ELSE
      BEGIN
         X2 := X * X;
         SUM := X;
         TERM := X;
         I := 0;
         REPEAT
            I := I + 1;
            SUM1 := SUM;
            TERM := 2.0 * TERM * X2/(1.0 + 2.0 * I);
            SUM := TERM + SUM1
         UNTIL TERM < TOL * SUM;
         ERF := 2.0 * SUM * EXP(−X2)/SQRTPI
      END (* IF *)
END (* erf *);

BEGIN (* main program *)
   DONE := FALSE;
   WRITELN;
```

Figure 11.5: An Infinite Series Expansion for the Gaussian Error Function (cont.)

```
REPEAT
   WRITE(' Arg? ');
   READLN(X);
   IF X < 0.0 THEN DONE : = TRUE
   ELSE
      BEGIN
         ANS : = ERF(X);
         WRITELN('Erf of ', X :6:3,
            ' is ', ANS :9:5)
      END
   UNTIL DONE
END.
```

Figure 11.5: An Infinite Series Expansion for the Gaussian Error Function (cont.)

Running the Program

The user must enter an argument that is equal to or larger than zero; the resulting function has a range from zero to unity. The result is zero if the argument is zero, and it approaches unity for arguments above 4. On the other hand, the function is approximately equal to its argument in the range of zero to 0.6. The error function has the same relative shape as the cumulative distribution function given in Figure 11.3. Selected values of the error function are given in Figure 11.6. The program will continually cycle, giving new values for the error function until a negative argument is entered.

y	$erf(y)$
0.0	0.0
0.1	0.1125
0.2	0.2227
0.3	0.3286
0.4	0.4284
0.5	0.5205
0.7	0.6778
1.0	0.8427
2.0	0.9953

Figure 11.6: The Gaussian Error Function

Suppose that zinc is to be diffused into a bar of copper at 900° C. We can use the error function to determine the concentration of copper as a function of time and the distance from the surface. The diffusion coefficient can be calculated from the diffusion constant and the activation energy. These latter quantities were found from the nonlinear curve fit in Chapter 10.

$$D = 0.31 \, e^{\frac{-45,420}{1.987(900+273)}}$$

If the distance is chosen to be 0.01 cm and the time is taken as 13 hours, then the argument y in the error function becomes 0.7. This corresponds to an error function of 0.67 and an *error function complement* of 0.33. This means that the concentration of zinc at a depth of 0.01 cm will be 33 percent of the surface concentration after 13 hours.

In the next section we will see that round-off errors make it difficult to find a direct solution of the error function complement. We will examine a program that uses two functions to handle both small and large arguments.

THE COMPLEMENT OF THE ERROR FUNCTION

For the above solution to the diffusion equation, we are actually interested in the complement of the error function rather than the error function itself. The complement is defined as:

$$\text{erfc}\,(y) = \frac{2}{\sqrt{\pi}} \int_{y}^{\infty} e^{-t^2} dt \tag{6}$$

The complement is obtained from the relationship:

$$\text{erfc}\,(y) = 1 - \text{erf}\,(y)$$

But if the complementary error function is always calculated in this way, there will be large round-off errors for arguments above 3. As erf approaches unity, $1 - \text{erf}$ approaches zero. Ultimately, all significant figures are lost. Furthermore, the computation time increases as the argument increases.

Alternatively, Equation 6 cannot be integrated by using the trapezoidal rule or Simpson's method. A problem occurs in selecting the upper limit for the integral. A value larger than 8 will produce a floating-point underflow because e^{-64} is so small. Yet the area from this point to infinity is significant and cannot be ignored.

PASCAL PROGRAM:
EVALUATING THE COMPLEMENT OF THE ERROR FUNCTION

One solution to this problem is to utilize two functions, one for small arguments and the other for larger arguments. The program given in Figure 11.7 utilizes this approach. The infinite-series expansion of erf, given in Figure 11.5, is used for smaller arguments. In addition, there is a second procedure for larger arguments: the complementary error function is calculated by means of an asymptotic expansion. This algorithm becomes more accurate as the argument increases. The equation, expressed as a continued fraction, rather than the usual infinite series, is:

$$\text{erfc}(y) = \frac{1/[1 + v/\{1 + 2v/[1 + 3v/(1 + \ldots)]\}]}{\sqrt{\pi}\, y e^{y^2}}$$

where

$$v = \frac{1}{2y^2}$$

```
PROGRAM ERFD3(INPUT, OUTPUT);

(* evaluate the Gaussian error function *)
(* Jan 30, 81 *)

VAR
    X, ER, EC : REAL;
    DONE : BOOLEAN;

FUNCTION ERF(X : REAL) : REAL;

(* Infinite series expansion of the
   Gaussian error function *)

CONST
    SQRTPI = 1.7724538;
    TOL = 1.0E−5;
```

Figure 11.7: The Error Function and its Complement

```
VAR
    X2, SUM, SUM1, TERM : REAL;
    I : INTEGER;
BEGIN
    X2 := X * X;
    SUM := X;
    TERM := X;
    I := 0;
    REPEAT
        I := I + 1;
        SUM1 := SUM;
        TERM := 2.0 * TERM * X2/(1.0 + 2.0 * I);
        SUM := TERM + SUM1
    UNTIL TERM < TOL * SUM;
    ERF := 2.0 * SUM * EXP(-X2)/SQRTPI
END (* erf *);

FUNCTION ERFC(X : REAL) : REAL;
(* Complement of error function *)

CONST
    SQRTPI = 1.7724538;
    TERMS = 12;

VAR
    X2, U, V, SUM : REAL;
    I : INTEGER;

BEGIN
    X2 := X * X;
    V := 1.0/(2.0 * X2);
    U := 1.0 + V * (TERMS + 1.0);
    FOR I := TERMS DOWNTO 1 DO
        BEGIN
            SUM := 1.0 + I * V/U;
            U := SUM
        END;
```

Figure 11.7: The Error Function and its Complement (cont.)

```
        ERFC := EXP(−X2)/(X * SUM * SQRTPI)
END (* ercf *);
BEGIN (* main program *)
    DONE := FALSE;
    WRITELN;
    REPEAT
        WRITE(' Arg? ');
        READLN(X);
        IF X < 0.0 THEN DONE := TRUE
        ELSE
            BEGIN
                IF X = 0.0 THEN
                    BEGIN
                        ER := 0.0;
                        EC := 1.0
                    END
                ELSE
                    BEGIN
                        IF X < 1.5 THEN
                            BEGIN
                                ER := ERF(X);
                                EC := 1.0 − ER
                            END
                        ELSE
                            BEGIN
                                EC := ERFC(X);
                                ER := 1.0 − EC
                            END (* IF *)
                    END;
                WRITELN(' X= ', X:6:2, ', Erf= ',
                    ER :7:4, ', Erfc= ', EC)
            END (* IF *)
    UNTIL DONE
END.
```

Figure 11.7: The Error Function and its Complement (cont.)

Running the Program

Type up the new version and execute it. The program will now print the error function and its complement. For arguments that are less than 1.5, the error function is calculated in function ERF using the infinite series expansion given in Figure 11.5. The complement is then obtained by subtraction from unity. On the other hand, if the argument is 1.5 or larger, then the complement is calculated from the asymptotic expansion using function ERFC. The error function is determined by subtraction from unity.

Compare the error function from this version with the data in Figure 11.6. Then try the values given in Figure 11.8 to check the complementary error function.

y	$erfc(y)$
1.5	$3.390E-2$
2.0	$4.678E-3$
2.5	$4.070E-4$
3.0	$2.209E-5$
3.5	$7.431E-7$
4.0	$1.542E-8$
4.5	$1.966E-10$

Figure 11.8: The Complementary Error Function

We will now modify our program so that it computes the error function using *nested parentheses* rather than a **FOR...DO** loop.

PASCAL PROGRAM:
A FASTER IMPLEMENTATION OF THE ERROR FUNCTION

In the previous program, the error function was calculated by summing successive terms in a loop. This iterative method of evaluation is slower than a direct summing of the terms. On the other hand, the iterative method will generally work with all commercial Pascal compilers. Nevertheless, a noniterative method is preferable if it can be implemented. Consequently, for the next version, we will write a variation in which the terms are summed directly.

The error function is calculated in the new version of function ERF as a polynomial of order 12. The coefficients are defined at the beginning of the function as T2, T3, T4, etc. A twelve-term polynomial was selected since this is the number of terms that are required for an argument of 1.5. Fewer terms are needed for smaller arguments. But, even so, this method is considerably faster than the previous iterative technique.

The complementary error function is also directly summed instead of being calculated iteratively as in the previous version. Again, the result is obtained more quickly.

There is a potential problem with the new version. Several Pascal compilers cannot process multiple levels of parentheses. As a consequence, the expressions for both the error function and its complement are separated into two or three parts. Even so, however, this will be too great a complexity for at least one Pascal compiler. For this reason, you may not be able to compile the new version. However, you should at least attempt to compile this version since it will run faster than the previous one. Make a copy of the previous program and alter the functions so that they look like Figure 11.9.

```
PROGRAM ERF4(INPUT, OUTPUT);
(* evaluation of the Gaussian error function *)
(* Jan 30, 81 *)
VAR
    X, ER, EC : REAL;
    DONE : BOOLEAN;
FUNCTION ERF(X : REAL) : REAL;
(* Infinite series expansion of the
    Gaussian error function *)
CONST
    SQRTPI = 1.7724538;
    T2 = 0.6666667;
    T3 = 0.2666667;
    T4 = 0.07619048;
    T5 = 0.01693122;
    T6 = 3.078403E − 3;
    T7 = 4.736005E − 4;
    T8 = 6.314673E − 5;
```

Figure 11.9: A Noniterative Error-Function Calculation

```
        T9 = 7.429027E−6;
        T10 = 7.820028E−7;
        T11 = 7.447646E−8;
        T12 = 6.476214E−9;

VAR
    X2, SUM : REAL;

BEGIN (* function erf *)
    X2 := X * X;
    SUM := T5 + X2 * (T6 + X2 * (T7 + X2 * (T8 + X2 * (T9
            + X2 * (T10 + X2 * (T11 + X2 * T12))))));
    ERF := 2.0 * EXP(−X2)/SQRTPI
            * (X * (1 + X2 * (T2 + X2 * (T3 + X2 * (T4 + X2
            * SUM)))))
END (* function erf *);

FUNCTION ERFC(X : REAL) : REAL;
(* error function complement *)

CONST
    SQRTPI = 1.7724538;

VAR
    X2, V, SUM : REAL;

BEGIN (* function erfc *)
    X2 := X * X;
    V := 1.0/(2.0 * X2);
    SUM := V/(1 + 8 * V/(1 + 9 * V/(1 + 10 * V
            / (1 + 11 * V/(1 + 12 * V)))));
    SUM := V/(1 + 3 * V/(1 + 4 * V/(1 + 5 * V
            / (1 + 6 * V/(1 + 7 * SUM)))));
    ERFC := 1.0/(EXP(X2) * X * SQRTPI
            * (1 + V/(1 + 2 * SUM)))
END (* function ercf *);
```

Figure 11.9: A Noniterative Error-Function Calculation (cont.)

```
BEGIN (* main program *)
   DONE := FALSE;
   WRITELN;
   REPEAT
      WRITE(' Arg? ');
      READLN(X);
      IF X < 0.0 THEN DONE := TRUE
      ELSE
         BEGIN
            IF X = 0.0 THEN
               BEGIN
                  ER := 0.0;
                  EC := 1.0
               END
            ELSE
            BEGIN
               IF X < 1.5 THEN
                  BEGIN
                     ER := ERF(X);
                     EC := 1.0 - ER
                  END
               ELSE
                  BEGIN
                     EC := ERFC(X);
                     ER := 1.0 - EC
                  END (* IF *)
            END;
            WRITELN(' X= ', X:6:2, ', Erf= ',
               ER :7:4, ', Erfc= ', EC)
         END (* IF *)
   UNTIL DONE
END.
```

Figure 11.9: A Noniterative Error-Function Calculation (cont.)

In the next section we will examine the special properties of the Gamma function and we will implement a program to evaluate it.

THE GAMMA FUNCTION

A function that is related to the error function is known as the Gamma function. It is defined by the integral:

$$\Gamma(n) = \int_0^\infty x^{n-1}e^{-x}dx \tag{7}$$

The Gamma function is important because it is part of the solution to Bessel's equation. In addition, it can be used to calculate factorials because of the recursive relationship:

$$\Gamma(n+1) = n\Gamma(n) \tag{8}$$

Since $\Gamma(1) = 1$, we can see that

$$
\begin{aligned}
\Gamma(2) &= \ \Gamma(1) = 1 &&= 1! \\
\Gamma(3) &= 2\Gamma(2) = 1\cdot 2 &&= 2! \\
\Gamma(4) &= 3\Gamma(3) = 1\cdot 2\cdot 3 &&= 3! \\
&\ \cdots \\
\Gamma(n) &= (n-1)\Gamma(n-1) = (n-1)!
\end{aligned}
$$

Thus, the general formula for using the Gamma function to calculate factorials is:

$$\Gamma(n+1) = n\Gamma(n) = n!$$

Since the Gamma function is defined for all real arguments greater than zero, the "factorial" of noninteger arguments can be defined as well. In addition, we find that:

$$\Gamma(0.5) = -0.5! = \sqrt{\pi}$$

This will be useful in calculating Bessel functions. The Gamma function is also defined for noninteger negative numbers. Its value, however, is infinite for zero and negative integers. A plot of the Gamma function for real arguments is given in Figure 11.10.

The Gamma function can be calculated using a form of Stirling's approximation:

$$\Gamma(x) = \sqrt{\frac{2\pi}{x}} \ x^x e^y$$

where

$$y = \frac{1}{12x} - \frac{1}{360x^3} - x$$

Since this is an asymptotic series, the relative accuracy improves as the argument increases.

PASCAL PROGRAM: EVALUATION OF THE GAMMA FUNCTION

The program given in Figure 11.12 uses Stirling's approximation to calculate the Gamma function. Arguments can be any real number greater than zero and less than about 32. The upper limit depends on the floating point arithmetic of your Pascal. Arguments can also be negative if they are nonintegral. However, the argument cannot be zero or a negative integer.

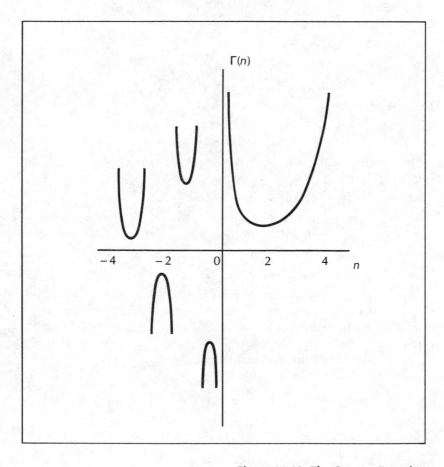

Figure 11.10: The Gamma Function

Positive arguments are incremented by 2, and the Gamma function of the new argument is calculated. The resulting Gamma function is then reduced to the corresponding original argument by using the algorithm:

$$\Gamma(x) = \frac{\Gamma(x+2)}{x(x+1)}$$

This conversion is not needed for larger arguments, but it insures that there will be at least six figures of precision for all values of x.

Negative arguments are incremented until they are positive, and the Gamma function is called recursively to calculate the new value. The result is then corrected to the original argument.

Running the Program

Type up the program and execute it. Try the values given in Figure 11.11 and compare the values for $\Gamma(x)$ with your results. The program cycles repeatedly. Enter a number smaller than -22 to terminate the program. Since:

$$x! = \Gamma(x+1)$$

the driver can be readily rewritten so that it will generate factorials rather than the Gamma function.

x	$\Gamma(x)$	
1	1	0!
2	1	1!
3	2	2!
4	6	3!
5	24	4!
6	120	5!
0.5	1.7725	$\sqrt{\pi}$
-0.5	-3.5449	$-\Gamma(0.5)/0.5$
-1.5	2.3633	$-\Gamma(-0.5)/1.5$

Figure 11.11: Selected Gamma Function Values

```
PROGRAM TSTGAM(INPUT, OUTPUT);
(* test the gamma function *)
(* Jan 26, 81 *)

VAR
   X : REAL;

FUNCTION GAMMA(X : REAL) : REAL;
 (* Jan 26, 81 *)

CONST
   PI = 3.1415926;

VAR
   I, J : INTEGER;
   Y, GAM : REAL;


BEGIN (* gamma function *)
   IF X >= 0.0 THEN
      BEGIN
         Y := X + 2.0;
         GAM := SQRT(2 * PI/Y)
            * EXP(Y * LN(Y) + (1 − 1/(30 * Y * Y))
            /(12 * Y) − Y);
         GAMMA := GAM/(X * (X + 1))
      END
   ELSE (* x < 0 *)
      BEGIN
         J := 0;
         Y := X;
         REPEAT (* increment argument until positive *)
            J := J + 1;
            Y := Y + 1.0
```

Figure 11.12: Evaluation of the Gamma Function

```
        UNTIL Y > 0.0;
        GAM := GAMMA(Y) (* recursive call *);
        FOR I := 0 TO J — 1 DO
          GAM := GAM/(X + I);
        GAMMA := GAM
      END (* x < 0 *)
  END (* gamma function *);

  BEGIN (* main program *)
    WRITELN;
    REPEAT
      REPEAT
        WRITE(' X: ');
        READLN(X)
      UNTIL X <> 0.0;
      WRITELN(' Gamma is ', GAMMA(X))
    UNTIL X < —22.0
  END.
```

Figure 11.12: Evaluation of the Gamma Function (cont.)

In the final two sections of this chapter we will introduce the Bessel functions of the first and second kind. *Bessel's equation* has many mathematical and scientific applications; solutions to this differential equation can be found using the Bessel functions. We will examine two Pascal implementations for these functions.

BESSEL FUNCTIONS

Bessel's equation:

$$x^2 y'' + xy' + (x^2 - n^2)y = 0$$

arises in the analysis of many different kinds of problems involving circular symmetry. In this equation, x is the independent variable, y is the dependent variable and n is a constant known as the order. This is a nonlinear differential equation that cannot be solved in closed form.

One of the solutions to Bessel's equation is:

$$y = J_n(x)$$

where J is the n-order Bessel function of the first kind.

The Bessel functions have been extensively tabulated for particular values of x and n. However, they are difficult to use in this form. For values of x less than about 15, the J Bessel functions can be calculated from the infinite series:

$$J_n(x) = \sum_{k=0}^{n} \frac{(-1)^k}{k!\Gamma(n+k+1)} \left(\frac{x}{2}\right)^{n+2k} \tag{9}$$

On the other hand, the asymptotic expression:

$$J_n(x) = \sqrt{\frac{2}{\pi x}} \cos\left(x - \frac{\pi}{4} - \frac{n\pi}{2}\right) \tag{10}$$

can be used for larger values of x. The Bessel functions J_0 and J_1 are shown in Figure 11.13.

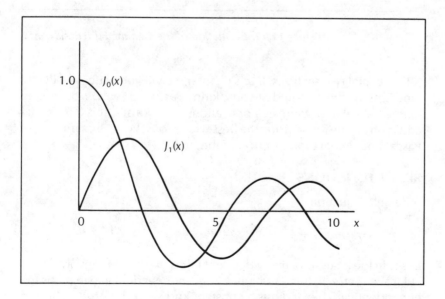

Figure 11.13: The Bessel Functions J_0 and J_1

PASCAL PROGRAM: BESSEL FUNCTIONS OF THE FIRST KIND

The program shown in Figure 11.15 uses both Equations 9 and 10 to calculate the Bessel functions of the first kind. The order can be zero or a positive number. The argument can also be a noninteger negative number. The Gamma function from the previous section is incorporated. The infinite series is utilized for arguments less than 15, and the asymptotic expression is used for greater arguments. There may be inaccuracies in this transition region depending on the floating-point arithmetic of your Pascal.

Type up the program and execute it. Try the values in Figure 11.14 for the order and argument, and compare the results. The program can be terminated by entering an order less than -25.

Since Bessel's equation is a second order equation, two independent solutions are needed. When the order n is not an integer, then both solutions can be obtained from the J Bessel functions. The expression is:

$$y = AJ_n(x) + BJ_{-n}(x)$$

where A and B are constants to be determined from the boundary conditions.

Order	Argument	Function
1	1	0.4401
0	1	0.7652
1	0.5	0.2423
0	0.5	0.9385
1	10	0.04347
0	10	-0.2459
0.25	1	0.7522
0.25	1.5	0.6192
0.5	1.5708 $(\pi/2)$	0.6366 $(2/\pi)$
-0.25	1	0.6694
-0.25	1.5	0.3180
-0.75	1.5	-0.2684

Figure 11.14: Selected Bessel Function Values

```
PROGRAM TSTBES(INPUT, OUTPUT);
(* test Bessel function *)
(* gamma function is included *)

VAR
   DONE : BOOLEAN;
   X, ORDR : REAL;

FUNCTION GAMMA(X : REAL) : REAL;
 (* Jan 26, 81 *)

CONST
   PI = 3.1415926;

VAR
   I, J : INTEGER;
   Y, GAM : REAL;


BEGIN (* gamma function *)
   IF X >= 0.0 THEN
      BEGIN
         Y := X + 2;
         GAM := SQRT(2 * PI/Y) * EXP(Y * LN(Y)
                  + (1 - 1/(30 * Y * Y))/(12 * Y) - Y);
         GAMMA := GAM/(X * (X + 1))
      END
   ELSE (* x < 0 *)
      BEGIN
         J := 0;
         Y := X;
         REPEAT
            J := J + 1;
            Y := Y + 1.0
```

Figure 11.15: The Bessel Function of the First Kind

```
        UNTIL Y > 0.0;
        GAM := GAMMA(Y);
        FOR I := 0 TO J − 1 DO
            GAM := GAM/(X + I);
        GAMMA := GAM
    END (* IF *)
END (* gamma function *);


FUNCTION BESSJ(X, N : REAL) : REAL;
(* cylindrical Bessel function *)
(* of the first kind *)
(* the gamma function is required *)
(* Feb 2, 81 *)


CONST
    TOL = 1.0E−5;
    PI = 3.1415926;
VAR
    I : INTEGER;
    TERM, NEW_TERM, SUM, X2 : REAL;


BEGIN (* bessj *)
    X2 := X * X;
    IF (X = 0.0) AND (N = 1.0) THEN BESSJ := 0.0
    ELSE IF X > 15 THEN (* asymptotic expansion *)
        BESSJ := SQRT(2/(PI * X)) * COS(X − PI/4 − N * PI/2)
    ELSE
        BEGIN (* regular infinite series *)
            IF N = 0.0 THEN SUM := 1.0
            ELSE SUM := EXP(N * LN(X/2))/GAMMA(N + 1.0);
            NEW_TERM := SUM;
            I := 0;
```

Figure 11.15: The Bessel Function of the First Kind (cont.)

```
          REPEAT
              I := I + 1;
              TERM := NEW _ TERM;
              NEW _ TERM := −TERM * X2 * 0.25/(I * (N + I));
              SUM := SUM + NEW _ TERM
          UNTIL ABS(NEW _ TERM) <= ABS(SUM * TOL);
          BESSJ := SUM
      END (* IF *)
  END (* bessj *);


  BEGIN (* main program *)
    DONE := FALSE;
    REPEAT
      WRITE(' Order: ');
      READLN (ORDR);
      IF ORDR < −25.0 THEN DONE := TRUE
      ELSE
        BEGIN
          WRITE(' X: ');
          READLN(X);
          WRITELN(' J Bessel is ', BESSJ(X, ORDR))
        END
    UNTIL DONE
  END.
```

Figure 11.15: The Bessel Function of the First Kind (cont.)

PASCAL PROGRAM: BESSEL FUNCTIONS OF THE SECOND KIND

If the order of Bessel's equation is an integer, then the solutions $J_n(x)$ and $J_{-n}(x)$ are linearly dependent. In this case, a second, independent solution can be obtained from the expression:

$$y = AJ_n(x) + BY_n(x)$$

where Y is a Bessel function of the second kind.

The program given in Figure 11.18 can be used to calculate Bessel functions of the second kind. While the order can be any real number, it is customary to use these functions only for integer orders. Two different algorithms are utilized. For arguments less than 12, the values of Y_0 and Y_1 are calculated from the expressions:

$$Y_0(x) = \frac{2}{\pi} \sum_{m=0}^{n} (-1)^m \left(\frac{x}{2}\right)^{2m} \frac{[\ln \frac{x}{2} + \gamma - h]}{(m!)^2}$$

and

$$Y_1(x) = -\frac{2}{\pi x} + \frac{2}{\pi} \sum_{m=1}^{n} (-1)^{m+1} \left(\frac{x}{2}\right)^{2m-1} \frac{\left[\ln\left(\frac{x}{2}\right) + \gamma - h + \frac{1}{2m}\right]}{(m!)(m-1)!}$$

where:

$$h = \sum_{r=1}^{m} \frac{1}{r} \quad \text{if } m \geq 1$$

and γ is Euler's constant ($\gamma = 0.57721566$). Orders other than 0 and 1 are calculated from the formula:

$$Y_n(x) = \frac{2n}{x} Y_{n-1}(x) - Y_{n-2}(x)$$

For larger arguments, an asymptotic expansion similar to the one used for the J Bessel functions is employed:

$$Y_n(x) = \sqrt{\frac{2}{\pi x}} \sin\left(x - \frac{\pi}{4} - \frac{n\pi}{2}\right)$$

A plot of the functions Y_0 and Y_1 is given in Figure 11.16.

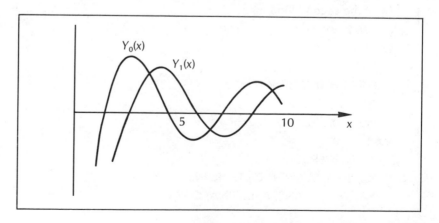

Figure 11.16: The Bessel Functions Y_0 and Y_1

Type up the program and try it out. The argument must be a positive number, since the function goes to minus infinity at zero. Some typical values are shown in Figure 11.17.

Argument	Y_0	Y_1
1	0.088	-0.781
2	0.510	-0.107
3	0.377	0.325
11	-0.169	0.164
15	0.206	0.021

Figure 11.17: Selected Values of the Bessel Function of the Second Kind

```
PROGRAM BESY(INPUT, OUTPUT);
(* evaluation of Bessel function *)
(* of the second kind *)
(* Feb 2, 81 *)
VAR
   X, ORDR : REAL;
   DONE : BOOLEAN;
FUNCTION BESSY(X, N : REAL) : REAL;
(* cylindrical Bessel function *)
(* of the second kind *)
(* Feb 2, 81 *)
CONST
   SMALL = 1.0E−8;
   EULER = 0.57721566;
   PI = 3.1415926;
   PI2 = 0.63661977 (* 2/pi *);
VAR
   J : INTEGER;
   X2, SUM, SUM2, T, T2, TS, TERM, XX, Y0, Y1,
   YA, YB, YC, ANS, A, B, SINA, COSA : REAL;
```

Figure 11.18: The Bessel Function of the Second Kind

```
BEGIN (* function bessy *)
   IF X < 12 THEN
     BEGIN
         XX := 0.5 * X;
         X2 := XX * XX;
         T := LN(XX) + EULER;
         SUM := 0.0;
         TERM := T;
         Y0 := T;
         J := 0;
         REPEAT
            J := J + 1;
            IF J <> 1 THEN SUM := SUM + 1/(J - 1);
            TS := T - SUM;
            TERM := -X2 * TERM/(J * J) * (1 - 1/(J * TS));
            Y0 := Y0 + TERM
         UNTIL ABS(TERM) < SMALL;
         TERM := XX * (T - 0.5);
         SUM := 0.0;
         Y1 := TERM;
         J := 1;
         REPEAT
            J := J + 1;
            SUM := SUM + 1/(J - 1);
            TS := T - SUM;
            TERM := (-X2 * TERM)/(J * (J - 1)) *
                   ((TS - 0.5/J)/(TS + 0.5/(J - 1)));
            Y1 := Y1 + TERM
         UNTIL ABS(TERM) < SMALL;
         Y0 := PI2 * Y0;
         Y1 := PI2 * (Y1 - 1/X);
         IF N = 0.0 THEN ANS := Y0
         ELSE IF N = 1.0 THEN ANS := Y1
         ELSE
```

Figure 11.18: The Bessel Function of the Second Kind (cont.)

```
          BEGIN (* find Y by recursion *)
              TS := 2.0/X;
              YA := Y0;
              YB := Y1;
              FOR J := 2 TO TRUNC(N + 0.01) DO
                BEGIN
                    YC := TS * (J — 1) * YB — YA;
                    YA := YB;
                    YB := YC
                END;
              ANS := YC
          END;
        BESSY := ANS;
     END (* x < 12 *)
  ELSE (* x > 11, asymptotic expansion *)
     BESSY := SQRT(2/(PI * X)) * SIN(X — PI/4 — N * PI/2)
END (* function bessy *);
BEGIN (* main program *)
  DONE := FALSE;
  WRITELN;
  REPEAT
    WRITE(' Order? ');
    READLN (ORDR);
    IF ORDR < 0.0 THEN DONE := TRUE
    ELSE
      BEGIN
        REPEAT
          WRITE(' Arg? ');
          READLN(X)
        UNTIL X >= 0.0;
        WRITELN(' Y Bessel is ', BESSY(X, ORDR))
      END (* IF *)
  UNTIL DONE
END.
```

Figure 11.18: The Bessel Function of the Second Kind (cont.)

The program can be terminated by entering a number less than zero.

SUMMARY

In Chapter 11 we have reviewed a number of the concepts and tools examined in this book. With the tools that we now have available to us we have found that we can implement some rather advanced mathematical applications. We saw Pascal programs that evaluate several variations of the Gaussian error function, the Gamma function, and the Bessel functions. In the process of developing these and other programs in this book, we have demonstrated the elegance of Pascal programming for technical applications.

APPENDIX **A**

Reserved Words and Functions

RESERVED WORDS

The following Pascal reserved words appear in **boldface** in this book:

AND	**ARRAY**	**BEGIN**	**CASE**	**CONST**
DIV	**DO**	**DOWNTO**	**ELSE**	**END**
EXTERN*	**FILE**	**FOR**	**FUNCTION**	**GOTO**
IF	**IN**	**LABEL**	**MOD**	**NIL**
NOT	**OF**	**OR**	**PACKED**	**PROCEDURE**
PROGRAM	**RECORD**	**REPEAT**	**SET**	**THEN**
TO	**TYPE**	**UNTIL**	**VAR**	**WHILE**
WITH				

*Implementation dependent

BUILT-IN FUNCTIONS

Name	*Action*
ABS	Absolute value
ARCTAN	Arc tangent
CHR	Convert integer to character

COS	Cosine
EOLN	True when end-of-line is reached
EOF	True when end-of-file is reached
EXP	e raised to the power
LN	Natural logarithm
ODD	True if the integer argument is odd
ORD	Convert character to integer
PRED	Predecessor or next lower (char, integer or scalar type)
ROUND	Convert real to integer by rounding
SIN	Sine
SQR	Square
SQRT	Square root
SUCC	Successor or next higher (char, integer or scalar type)
TRUNC	Convert real to integer by truncation

APPENDIX **B**

Summary of Pascal

MINIMUM STANDARD CHARACTER SET

A full alphabet	A - Z *and/or* a - z
The digits	0 - 9
The special characters	+ – * / = < > ( ) [ ] . , ; : ↑
A space or blank character	

VARIABLE NAMES

A Pascal variable name may be any contiguous sequence of alphabetic and numeric characters beginning with a letter. Some implementations allow certain non-alphanumeric characters, such as the underscore character.

Examples:

SEED   LINE1   SUM_X2   Y_CALC

The maximum *distinctive* name length varies from one Pascal compiler to the next. Some are limited to a maximum of 6 or 8 characters, while others allow names of 32 characters or more. Thus, in this book identifiers such as:

SUM_X_SQUARED

and

SUM_Y_SQUARED

have been used, rather than:

SUM_SQUARE_X

and

SUM_SQUARE_Y

NUMBERS

Integers are signed or unsigned strings of digits.

Examples:

    15 −19253 +7 0

Reals are written with a decimal point, a scale factor or both:

    379.1275 3.791275E2 3791275E-4

The E notation indicates multiplication by powers of 10. Thus, E2 signifies 10^2. E must always be followed by an integer, signed or unsigned.

COMMENTS

Comments are typed between the symbols (* and *), and are ignored by the compiler.

Example:

    (*get values for n and arrays x, y*)

On some compilers, the standard symbols { and } may be used as comment delimiters.

OPERATIONS

Integer Operations

+	Addition
−	Subtraction
*	Multiplication
DIV	Division (truncated to integer)
MOD	Modulo (supplies the *remainder* from the division of two integers)

Real Operations

+	Addition
−	Subtraction
*	Multiplication
/	Division

Boolean Operations

AND OR NOT

Relational Operations

These operations result in Boolean values TRUE or FALSE.

<	Less than
>	Greater than
=	Equal to
<=	Less than or equal to
>=	Greater than or equal to
<>	Not equal to
IN	Used with data type **SET**, to determine membership of an element (See "Set Types".)

SYNTAX

Program Heading

> **PROGRAM** PROGRAM_NAME (FILE_NAME, FILE_NAME,...);

Example:

> **PROGRAM** TSTSORT (INPUT,OUTPUT);

Constant Definition

> **CONST** CONST_NAME = value; CONST_NAME = value;...

Example:

> **CONST**
> MAXR = 20; (*data points*)
> MAXC = 4; (*polynomial terms*)

There are some predefined constants in Pascal:

TRUE	Boolean true value
FALSE	Boolean false value
MAXINT	Largest integer the computer can work with
NIL	Null pointer

Note that a constant must be defined as a simple value, *not* an expression. The second statement below is illegal:

```
CONST
    Y = 3; (*legal*)
    X = Y + 1 (*illegal*)
```

Variable Definition

```
VAR
    VAR_NAME, VAR_NAME,...:TYPE;
    VAR_NAME, VAR_NAME,...:TYPE;...
```

Example:

```
VAR
    X, X2: REAL;
    ALLDONE: BOOLEAN;
    ERROR: BOOLEAN;
```

Assignment Statements

```
VAR_NAME := expression
```

Example:

```
X := SEED + PI;
X := EXP(5.0*LN(X));
SEED := X − TRUNC(X);
RANDOM := SEED
```

The Compound Statement

A compound statement is any sequence of statements bracketed by a **BEGIN/END** pair.

Example:

```
BEGIN(*swap*)
    HOLD := P;
    P := Q;
    Q := HOLD
END(*swap*);
```

Procedure Definition

PROCEDURE PROC _ NAME (value parameters; **VAR** variable parameters); body of procedure

Example:

PROCEDURE DERIV (X: REAL; **VAR** FX, DFX: REAL);

VAR
    E : REAL;

BEGIN(*deriv*)
    E := EXP(X);
    FX := E − 4.0*X;
    DFX := E − 4.0
END(*deriv*);

Function Definition

FUNCTION FUNC _ NAME (value parameters): result _ type; body of function

The value of a function result is assigned by a statement (within the body of the function) of the form

FUNCT _ NAME := value;

Example:

FUNCTION DETER (A : ARY2) : REAL;
(*Calculate the determinant of a 3-by-3 matrix*)
VAR
    SUM : REAL;
BEGIN(*deter*)
    SUM := A[1,1] * (A[2,2] * A[3,3] − A[3,2] * A[2,3])
            − A[1,2] * (A[2,1] * A[3,3] − A[3,1] * A[2,3])
            + A[1,3] * (A[2,1] * A[3,2] − A[3,1] * A[2,2]);
    DETER := SUM
END(*deter*);

Placement of Semicolons

Semicolons are required between any two Pascal statements. They are not required either immediately after a **BEGIN** or immediately before an **END**, but their presence there is harmless. A semicolon must *never* immediately precede the **ELSE** in an **IF-THEN-ELSE** construction.

CONDITIONAL STATEMENTS

The IF-THEN Statement

IF expression **THEN** statement

Example:

IF B[N,N] = 0.0 **THEN** ERROR := TRUE

The IF-THEN-ELSE Statement

IF expression **THEN** statement **ELSE** statement

Example:

```
IF DET = 0.0 THEN
  BEGIN
    ERROR := TRUE;
    WRITELN('ERROR: matrix singular')
  END
ELSE
  BEGIN
    SETUP(B,COEF,1);
    SETUP(B,COEF,2);
    SETUP(B,COEF,3)
  END(*ELSE*)
```

The CASE Statement

```
CASE expression OF
    case-labels:statement;
    case-labels:statement;
            .
            .
            .
    case-labels:statement
END
```

Example:

```
CASE YEAR OF
    1: WRITE('FRESHMAN');
    2: WRITE('SOPHOMORE');
    3: WRITE('JUNIOR');
    4: WRITE('SENIOR');
  5,6,7:WRITE('GRADUATE')
  END;
```

ITERATIVE STATEMENTS

The WHILE-DO Statement

```
WHILE expression DO statement
```

Example:

```
WHILE PIVOT < A[J] DO J := J — 1;
```

The REPEAT-UNTIL Statement

```
REPEAT group-of-statements UNTIL expression
```

Example:

```
REPEAT
    WRITE('How many equations? ');
    READLN(N)
UNTIL N < MAXR;
```

The FOR-TO-DO Statement

FOR variable : = first-value **TO** last-value **DO** statement

Example:

FOR J : = 1 **TO** N **DO** WRITE (COEF[J]);

The FOR-DOWNTO-DO statement

FOR variable : = first-value **DOWNTO** last-value **DO** statement

Example:

```
FOR I := TERMS DOWNTO 1 DO
  BEGIN
    SUM := 1.0 + I * V/U
    U := SUM
  END;
```

TRANSFER-OF-CONTROL STATEMENTS

It is recommended that **GOTO** statements be avoided except in the rare cases where they actually have some unique value.

Label Declaration

If labels are to be used in a Pascal program they must be explicitly predeclared immediately after the CONST declaration section.

LABEL integer, integer,...;

Example:

LABEL 99;

GOTO Statement _____

> **GOTO** label;

Example:

> **BEGIN** (∗Procedure gaussj∗)
>   **IF** ERROR **THEN GOTO** 99;
>   .
>   .
>   .
>
>   99:
>   **END** (∗Procedure gaussj∗);

The **GOTO** in this example causes a clear and orderly termination of the procedure if ERROR is true. The alternative, **IF NOT** ERROR **THEN**, could be harder to manage. Using a **GOTO** statement to jump out of a procedure is questionable programming practice, although some compilers allow it.

INPUT AND OUTPUT

Input Procedures _____

> READ(variables);
> READLN(variables);
> READ(FILE _ NAME,variables);
> READLN(FILE _ NAME,variables);

FILE _ NAME must be a variable explicitly declared as of type **FILE** or passed as a parameter in the program heading. Note that type TEXT is equivalent to **FILE OF CHAR**.

A READ statement gets values from the current line (strictly speaking, "record"), and assigns them to its variables, leaving any values remaining on the line available for subsequent input.

A READLN statement inputs to its variables from the current line, passing over whatever else may remain on that line. Subsequent input starts from the beginning of the next line.

Output Procedures

```
WRITE(expressions);
WRITELN(expressions);
WRITE(FILE_NAME,expressions);
WRITELN(FILE_NAME,expressions);
```

A WRITE statement simply appends its output to the current output line or record. A WRITELN statement appends its output to the current line, and then issues a carriage return/line feed, i.e., terminates the record.

Examples:

```
WRITE(' X: ');
READLN(X);
WRITELN(' GAMMA IS ', GAMMA(X));
```

Formatting Numeric Output

Each expression in a WRITE or WRITELN can be followed by one or two integers in the following form:

```
integer_expression:integer
real_expression:integer:integer
```

The value of "expression" is output. The first integer is the character width of the output field. The actual value output is right adjusted in this field and preceded by blanks. All the digits of the value will always be output even if this output overflows the field width.

In the case of a real number, the second integer defines the number of digits to be output to the right of the decimal point.

Example:

```
WRITELN(I:3, X[I]:8:1, Y[I]:9:2, Y_CALC[I]:9:2);
```

DATA TYPES

TYPE declarations follow the **CONST** declarations and precede the **VAR** declarations.

Scalar Types

TYPE TYPE _ NAME = (identifier, identifier,...);

Example:

TYPE
DAY _ OF _ WEEK = (MONDAY, TUESDAY, WEDNESDAY,
THURSDAY, FRIDAY,
SATURDAY, SUNDAY);

Subrange Types

TYPE
TYPE _ NAME = constant..constant;
VAR
VAR _ NAMES : constant..constant;

Example:

TYPE
INDEX = 1..MAX;
WEEK _ DAY = (MONDAY..FRIDAY);

Array Types

TYPE
TYPE _ NAME = **ARRAY**[index _ type,index _ type,...]
OF type;
VAR
VAR _ NAMES : **ARRAY**[index _ type, index _ type,...]
OF type;

A multidimensional array requires that an index _ type be specified for each dimension. An index _ type is usually a subrange type but may be any scalar type.

Examples:

```
TYPE
    ARYS  = ARRAY[1..CMAX] OF REAL;
    ARY2S = ARRAY[1..RMAX, 1..CMAX] OF REAL;
            (*TWO DIMENSIONAL*)

VAR
    W      : ARRAY[1..MAXC, 1..MAXC] OF REAL;
             (*MAXC BY MAXC*)
    INDEX : ARRAY[1..MAXC, 1..3] OF INTEGER;
            (*MAXC BY 3*)
```

Referencing Array Elements

Array indices appear between square brackets. In the case of multi-dimensional arrays, the index expressions are separated by commas. The subscript index for each dimension may be any expression that evaluates to the declared index type.

```
    ARRAY_NAME [index,index,...]
```

Example:

```
    C[I,J] := A[I,K] * B[K,J]
```

Packed Arrays

In some implementations the amount of memory space used by an array may be reduced by declaring the array as **PACKED**. The actual amount of space saved and the resulting slowdown in access time depend on the implementation.

```
TYPE
    TYPE_NAME = PACKED ARRAY [index_type,
                    index_type...]OF element_type;
VAR
    VAR_NAME: PACKED ARRAY [index_type,
                index_type...] OF element_type;
```

Example:

```
TYPE
    LINE_IMAGE = PACKED ARRAY [1...MAXCOL] OF CHAR
```

Record Types

```
TYPE
    TYPE_NAME = RECORD field-list END;
VAR
    VAR_NAME: RECORD field-list END;
```

Examples:

```
TYPE
    COMPLEX = RECORD RE, IM: REAL END;
TYPE
    CUSTOMER = RECORD
        NAME,STREET:PACKED ARRAY[1..30] OF CHAR:
        CITY:PACKED ARRAY[1..20] OF CHAR;
        STATE:PACKED ARRAY[1..2] OF CHAR;
        ZIP: 0..99999
    END(*customer record*);
```

The fields of *variant* records may change from record to record. The different structures that may occur in the variant portion of the record are specified using a **CASE**-like construction that *must* follow the declaration of the *fixed* fields of the record.

Example:

```
TYPE
    DATE = RECORD
        MO: 1..12;
        YR: INTEGER
    END(*date*);
    GRADE = (FRESH, SOPH, JR, SEN);

    STUDENT = RECORD
        NAME: PACKED ARRAY [1..20] OF CHAR;
        NUM: INTEGER;
        CASE GR:GRADE OF
            FRESH, SOPH: (MAJOR: BOOLEAN);
            JUN: (CREDITS: INTEGER);
            SEN: (GDATE: DATE)
    END(*student*);
```

Referencing Record Fields

A field within a record variable is referenced by the so-called "dot" notation; the field name (which appears in the record type declaration) is joined with a dot to the actual record variable name:

    RECORD _ NAME. FIELD _ NAME

Examples:

        VAR
            COED: STUDENT; (∗ See type STUDENT above ∗)
            ...
        COED.NAME : = 'JANE SMITH
        COED.NUM : = 75210;
        COED.GR : = SEN;
        COED.GDATE.MO : = 6;
        COED.GDATE.YR : = 1981;
        ...

Where many accessses are to be made to fields within the same record variable(s) a context may be established using the **WITH** statement:

        WITH RECORD _ NAME **DO** statement

Example:

        VAR
            CUR _ ACCOUNT: CUSTOMER; (∗ See type CUSTOMER above ∗)
            ...
        WITH CUR _ ACCOUNT **DO**
            BEGIN
                READLN(NAME);
                READLN(STREET);
                READLN(CITY,STATE,ZIP)
            END;

The **WITH** construct makes the program clearer to read and easier to compile efficiently.

Set Types

```
TYPE
    TYPE_NAME = SET OF base-type;
VAR
    VAR_NAME: SET OF base-type;
```

Example:

```
TYPE
    DAYS = (MO, TU, WE, TH, FR, SA, SU);
    WEEK = SET OF DAYS;
VAR
    WORKDAY, HOLIDAY, WEEKDAY: WEEK;
```

Set Operations

+	Union
*	Intersection
—	Difference
=	Set equality
<>	Set inequality
IN	Set membership. TRUE if the scalar operand on the left is an element of the set operand on the right.
<=	Set inclusion. TRUE if all elements IN the left operand are also IN the right operand.
>=	Set containment. TRUE if all elements IN the right operand are also IN the left operand.

File Types

```
TYPE
    filename = FILE OF type;
VAR
    varname-list: FILE OF type;
```

Example:

```
VAR
    FILEA, FILEB, FILEC : FILE OF INTEGER;
```

The standard I/O procedures for use with files are:

RESET	Opens a file so that it can be read.
REWRITE	Opens a file so that it can be written (all components previously written on the file are lost).
GET	Transfers one component of a file to the associated buffer variable.
PUT	Appends the contents of a buffer variable to its file.
EOF	A Boolean function that is TRUE when no record is available under the file window.

Pointer Types

A pointer variable contains the address of a *dynamic variable*. Dynamic variables are variables that are created during program execution by the procedure NEW. Such variables are referenced indirectly by pointer variables. A pointer variable can be initialized to the value **NIL**, meaning the pointer is not pointing to any dynamic variable.

```
TYPE
    type _ name = type;
VAR
    var _ name: type;
```

Example: (A linked list)

```
TYPE
    LINK = PART; (*LINK IS A POINTER TO AN ITEM OF TYPE PART*)
    ...
    PART = RECORD
    ...
        NEXT: LINK; (*POINTS TO THE NEXT PART*)
        ...
    END (*PART RECORD*);
```

Bibliography

I. Pascal

Alagic, Suad, and Arbib, Michael. *The Design of Well-Structured and Correct Programs*. New York, Heidelberg, and Berlin: Springer-Verlag, 1978.

Cherry, George. *Pascal Programming Structures: An Introduction to Systematic Programming*. Reston, Va.: Reston, 1980.

Conway, Richard, and Archer, J. *Programming for Poets: A Gentle Introduction to Pascal*. Cambridge, Mass.: Winthrop Publishers, 1980.

Grogono, Peter. *Programming in Pascal*. Revised Edition. Reading, Mass.: Addison-Wesley, 1980.

Jensen, Kathleen, and Wirth, Niklaus. *Pascal User Manual and Report*. Second Edition. New York, Heidelberg, and Berlin: Springer-Verlag, 1980.

Ledgard, Henry; Heuras, John; and Nagin, Paul. *Pascal with Style: Programming Proverbs*. Rochelle Park, N.J.: Hayden, 1979.

Schneider, G. Michael; Weingart, Steven; and Perlman, David. *An Introduction to Programming and Problem Solving with Pascal*. New York: John Wiley & Sons, 1978.

Zaks, Rodnay. *Introduction to Pascal, Including UCSD Pascal*. Berkeley: Sybex, 1980.

II. General

Daniel, Cuthbert; Wood, Fred; and Gorman, John. *Fitting Equations to Data*. New York: John Wiley & Sons, 1971.

Fike, C.T. *Computer Evaluation of Mathematical Functions*. Englewood Cliffs, N.J.: Prentice-Hall, 1968.

Forsythe, George; Malcolm, Michael; and Moler, Cleve. *Computer Methods for Mathematical Computations*. Englewood Cliffs, N.J.: Prentice-Hall, 1977.

Fox, L., and Mayers, D.F. *Computing Methods for Scientists and Engineers*. New York: Oxford University Press, 1968.

Gilder, Jules. *BASIC Computer Programs in Science and Engineering*. Rochelle Park, N.J.: Hayden, 1980.

Hart, John F., et al. *Computer Approximations*. New York: John Wiley & Sons, 1968.

Hastings, Cecil, Jr. *Approximations for Digital Computers*. Princeton, N.J.: University Press, 1955.

Hewlett-Packard. *HP-25 Applications Programs*. Cupertino, Calif., 1975.

Hornbeck, Robert. *Numerical Methods*. New York: Quantum Publishers, 1975.

Hubin, Wilbert. *BASIC Programming for Scientists and Engineers*. Englewood Cliffs, N.J.: Prentice-Hall, 1978.

International Business Machines Corporation. *System/360 Scientific Subroutine Package, Programmer's Manual*. White Plains, N.Y., 1966.

Jennings, Alan. *Matrix Computation for Engineers and Scientists*. New York: John Wiley & Sons, 1977.

Kernighan, Brian W., and Plauger, P.J. *Software Tools*. Reading, Mass.: Addison-Wesley, 1976.

Khabaza, I.M. *Numerical Analysis*. Elmsford, N.Y.: Pergamon Press, 1965.

Kreyszig, Erwin. *Advanced Engineering Mathematics*. New York: John Wiley & Sons, 1967.

Ley, B. James. *Computer Aided Analysis and Design for Electrical Engineers*. New York: Holt, Rinehart and Winston, 1970.

McCormick, John, and Salvadori, Mario. *Numerical Methods in FORTRAN*. Englewood Cliffs, N.J.: Prentice-Hall, 1964.

Ruckdeshel, F.R. *BASIC Scientific Subroutines*. Volume 1. Peterborough, N.H.: Byte/McGraw-Hill, 1981.

Scheaffer, R.L., and Mendenhall, W. *Introduction to Probability: Theory and Applications*. N. Scituate, Mass: Duxbury Press, 1975.

Smith, J. *Advanced Analysis with the Sharp 5100 Scientific Calculator*. New York: John Wiley & Sons, 1979.

Sokolnikoff, Ivan, and Sokolnikoff, Elizabeth. *Higher Mathematics for Engineers and Physicists*. New York: McGraw-Hill, 1941.

Vandergraft, J.S. *Introduction to Numerical Computations*. New York: Academic Press, 1978.

Walpole, Ronald, and Myers, Raymond. *Probability and Statistics for Engineers and Scientists*. New York: Macmillan Co., 1972.

Wylie, Clarence, Jr. *Advanced Engineering Mathematics*. Second Edition. New York: McGraw-Hill, 1960.

Index

The SYBEX Library

INTRODUCTION TO COMPUTERS

DON'T (or How to Care for Your Computer)
by Rodnay Zaks   214 pp., 100 illustr., Ref. 0-065

The correct way to handle and care for all elements of a computer system, including what to do when something doesn't work.

YOUR FIRST COMPUTER
by Rodnay Zaks   258 pp., 150 illustr., Ref. 0-045

The most popular introduction to small computers and their peripherals: what they do and how to buy one.

FROM CHIPS TO SYSTEMS:
AN INTRODUCTION TO MICROPROCESSORS
by Rodnay Zaks   552 pp., 400 illustr., Ref. 0-063

A simple and comprehensive introduction to microprocessors from both a hardware and software standpoint: what they are, how they operate, how to assemble them into a complete system.

FOR YOUR APPLE

APPLE II® BASIC PROGRAMS IN MINUTES
by Stanley R. Trost   150 pp., illustr., Ref. 0-121

A collection of ready-to-run programs for financial calculations, investment analysis, record keeping, and many more home and office applications. These programs can be entered on your Apple II plus or IIe in minutes!

YOUR FIRST APPLE II® PROGRAM
by Rodnay Zaks   150 pp. illustr., Ref. 0-136

A fully illustrated, easy-to-use introduction to APPLE BASIC programming. Will have the reader programming in a matter of hours.

THE APPLE CONNECTION
by James W. Coffron   264 pp., 120 illustr., Ref. 0-085

Teaches elementary interfacing and BASIC programming of the Apple for connection to external devices and household appliances.

FOR YOUR IBM PC

THE BEST OF IBM® PC SOFTWARE
by Stanley R. Trost   144 pp., illustr., Ref. 0-104

Separates the wheat from the chaff in the world of IBM PC software. Tells you what to expect from the best available IBM PC programs.

IBM® PC DOS HANDBOOK
by Richard King   144 pp., illustr., Ref. 0-103
Explains the PC disk operating system, giving the user better control over the system. Get the most out of your PC by adapting its capabilities to your specific needs.

BUSINESS GRAPHICS FOR THE IBM® PC
by Nelson Ford   200 pp., illustr., Ref. 0-124
Ready-to-run programs for creating line graphs, complex illustrative multiple bar graphs, picture graphs, and more. An ideal way to use your PC's business capabilities!

THE IBM® PC CONNECTION
by James W. Coffron   200 pp., illustr., Ref. 0-127
Teaches elementary interfacing and BASIC programming of the IBM PC for connection to external devices and household appliances.

BUSINESS & PROFESSIONAL

INTRODUCTION TO WORD PROCESSING
by Hal Glatzer   205 pp., 140 illustr., Ref. 0-076
Explains in plain language what a word processor can do, how it improves productivity, how to use a word processor and how to buy one wisely.

MASTERING VISICALC®
by Douglas Hergert   217 pp., 140 illustr., Ref. 0-090
Explains how to use the VisiCalc "electronic spreadsheet" functions and provides examples of each. Makes using this powerful program simple.

VISICALC® FOR SCIENCE AND ENGINEERING
by Stanley R. Trost & Charles Pomernacki   225 pp., illustr., Ref. 0-096
More than 50 programs for solving technical problems in the science and engineering fields. Applications range from math and statistics to electrical and electronic engineering.

BASIC

YOUR FIRST BASIC PROGRAM
by Rodnay Zaks   150pp. illustr. in color, Ref. 0-129
A "how-to-program" book for the first time computer user, aged 8 to 88.

FIFTY BASIC EXERCISES
by J. P. Lamoitier   232 pp., 90 illustr., Ref. 0-056
Teaches BASIC by actual practice, using graduated exercises drawn from everyday applications. All programs written in Microsoft BASIC.

BASIC PROGRAMS FOR SCIENTISTS AND ENGINEERS
by Alan R. Miller   318 pp., 120 illustr., Ref. 0-073
This book from the "Programs for Scientists and Engineers" series provides a library of problem-solving programs while developing proficiency in BASIC.